T0300013

Chapman & Hall/CRC Mathematical Biology and Medicine Series

THE TEN MOST WANTED SOLUTIONS IN PROTEIN BIOINFORMATICS

CHAPMAN & HALL/CRC
Mathematical Biology and Medicine Series

Aims and scope:
This series aims to capture new developments and summarize what is known over the whole spectrum of mathematical and computational biology and medicine. It seeks to encourage the integration of mathematical, statistical and computational methods into biology by publishing a broad range of textbooks, reference works and handbooks. The titles included in the series are meant to appeal to students, researchers and professionals in the mathematical, statistical and computational sciences, fundamental biology and bioengineering, as well as interdisciplinary researchers involved in the field. The inclusion of concrete examples and applications, and programming techniques and examples, is highly encouraged.

Series Editors

Alison M. Etheridge
Department of Statistics
University of Oxford

Louis J. Gross
Department of Ecology and Evolutionary Biology
University of Tennessee

Suzanne Lenhart
Department of Mathematics
University of Tennessee

Philip K. Maini
Mathematical Institute
University of Oxford

Hershel M. Safer
Informatics Department
Zetiq Technologies, Ltd.

Eberhard O. Voit
Department of Biometry and Epidemiology
Medical University of South Carolina

Proposals for the series should be submitted to one of the series editors above or directly to:
CRC Press UK
23 Blades Court
Deodar Road
London SW15 2NU
UK

Chapman & Hall/CRC Mathematical Biology and Medicine Series

THE TEN MOST WANTED SOLUTIONS IN PROTEIN BIOINFORMATICS

ANNA TRAMONTANO

Chapman & Hall/CRC
Taylor & Francis Group
Boca Raton London New York Singapore

Published in 2005 by
CRC Press
Taylor & Francis Group
6000 Broken Sound Parkway NW, Suite 300
Boca Raton, FL 33487-2742

Library of Congress Cataloging-in-Publication Data

Tramontano, Anna.
   The ten most wanted solutions in protein bioinformatics / Anna Tramontano.
      p. cm. -- (Chapman & Hall/CRC mathematical biology and medicine series)
   Includes bibliographical references and index.
   ISBN 1-58488-491-6 (alk. paper)
   1. Proteomics. 2. Bioinformatics. I. Title. II. Series.

QP551.T723 2005
572'.6--dc22
                                                 2005041404

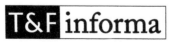

Taylor & Francis Group
is the Academic Division of T&F Informa plc.

Visit the Taylor & Francis Web site at
http://www.taylorandfrancis.com

and the CRC Press Web site at
http://www.crcpress.com

Foreword

The goal of protein bioinformatics is to assist experimental biology in assigning a function or suggesting functional hypotheses for all known proteins. The task is formidable. A simple calculation shows that we cannot possibly study each and every biological molecule of the universe. Therefore, we need fast and reliable computational methods to extrapolate the knowledge accumulated on a subset of cases to the rest of the protein universe.

This book reviews available methods in protein bioinformatics, with a special emphasis on their effectiveness in inferring the biological properties and functional roles of proteins. It is organized around specific problems that elicit the efforts of the community, and it focuses on the limitations of current approaches and on future developments that are likely to improve our understanding of the exquisitely specific and efficient mechanisms of protein function.

Bioinformatics is an interdisciplinary science that synergistically utilizes the contributions of informatics, physics, and mathematics, but, ultimately, the objective is the solution of biological problems. Therefore, this book starts with an overview of what we know about the structure and function of proteins. Proteins are a product of evolution. Thus, the basic principles of evolution must be kept in mind when new methods are devised or new routes are explored for inferring the function of a biological macromolecule. This book addresses the problem of detecting the existence of an evolutionary relationship between proteins in Problem 1. The detection of local similarities between protein sequences and the analysis of high-throughput experiments can also be effectively exploited for functional assignment, as is shown in Problems 2 and 3. Much more information can be derived from the knowledge of the three-dimensional structures of proteins. These structures can be experimentally determined or inferred from computational methods (Problems 4 and 5) and studied for insight into the roles of proteins (Problem 6). Proteins interact with each other and with ligands, both physically and logically (Problems 7 and 8), as parts of complex regulative networks. Several methods are being devised to explore these aspects of protein function. Finally, we discuss the extent to which our understanding of proteins allows us to design completely new proteins tailored to specific tasks (Problem 9) or to rationally modify the function and properties of existing proteins (Problem 10).

As we will see, many unsolved problems remain in each of these areas, and new ideas are continuously being produced and tested. The pressure on this relatively new discipline is strong because an understanding of life, in all its beauty and complexity, finally seems within our reach, and our astonishment at being so close to our goal is only equaled by our impatience to reach it.

Introduction

Proteins are the major components of living organisms and constitute more than 25% by weight of a typical cell. Even more impressive is the variety of functions that they can perform: catalysis, immune recognition, cell adhesion, signal transduction, sensory capabilities, transport, movement, and cellular organization. From a chemical perspective, proteins are linearly-oriented heteropolymers of amino acids (small organic molecules) whose structures and properties are described in this book. The sequence of amino acids in a protein is determined by the sequence of nucleotides, or bases, in the corresponding gene. Each adjacent triplet of bases of a gene in the DNA codes for one amino acid or for a codon that signals the end of the gene, according to the practically universal genetic code shown in Table 1.

The nucleotide sequence of a genomic region is technically much easier and faster to obtain than the sequence of the encoded protein, as is evidenced by the pace at which the complete genomes of many organisms, including *Homo sapiens*, are being deciphered. The large majority of known protein sequences are in fact deduced from the corresponding sequences of the genes, rather than from direct chemical sequencing of the proteins.

TABLE 1

The Genetic Code

First Base		Second Base				Third Base
		U	C	A	G	
	U	Phe (F)	Ser (S)	Tyr (Y)	Cys (C)	U
		Phe (F)	Ser (S)	Tyr (Y)	Cys (C)	C
		Leu (L)	Ser (S)	Stop	Stop	A
		Leu (L)	Ser (S)	Stop	Trp (W)	G
	C	Leu (L)	Pro (P)	His (H)	Arg (R)	U
		Leu (L)	Pro (P)	His (H)	Arg (R)	C
		Leu (L)	Pro (P)	Gln (Q)	Arg (R)	A
		Leu (L)	Pro (P)	Gln (Q)	Arg (R)	G
	A	Ile (I)	Thr (T)	Asn (N)	Ser (S)	U
		Ile (I)	Thr (T)	Asn (N)	Ser (S)	C
		Ile (I)	Thr (T)	His (H)	Arg (R)	A
		Met (M)	Thr (T)	His (H)	Arg (R)	G
	G	Val (V)	Ala (A)	Asp (D)	Gly (G)	U
		Val (V)	Ala (A)	Asp (D)	Gly (G)	C
		Val (V)	Ala (A)	Glu (E)	Gly (G)	A
		Val (V)	Ala (A)	Glu (E)	Gly (G)	G

Note: Each triplet of bases in a gene codes for one of the 20 amino acids, here listed in their three-letter and one-letter codes.

However, the genetic material not only contains genes, but also contains regulatory regions and noncoding regions of unknown function, such as long and short repeats; pseudogenes and retropseudogenes; satellite, mini-satellite, and microsatellite regions; transposons and retrotransposons; viral vestigials; and others. In higher organisms, the gene sequence is also interrupted by noncoding fragments of variable length called introns.

Although the large body of available genetic information holds the promise of unraveling the meaning of life, we must decode this information; that is, we must detect which regions are the gene-coding regions, translate these regions into the corresponding protein sequence, and work out the protein's molecular function. The development of methods for finding the genes and their corresponding proteins and for unraveling their function is essential. It is the only route to utilizing our genomic knowledge for rationally interfering with diseases and understanding, for example, the genetic basis of individual pharmacological responses.

In this book, we do not discuss the problem of finding genes, which is a major challenge that the genomic era is posing to bioinformatics. Rather, we concentrate on the techniques that can be applied to derive functional knowledge of a protein, once the complete sequence of its amino acids is known.

The Structure of Proteins

The function of a protein depends upon its "shape;" that is, upon the three-dimensional structure that can be determined by X-ray crystallography or nuclear magnetic resonance experiments. The resulting data are stored in a data base called the Protein Data Bank (PDB). At present the PDB contains a few thousand examples of protein structures, but it is rather redundant. Often, different examples of the structure of a protein have the same amino acid sequence but in different states, such as with different bound ligands, in complex with different proteins, or determined under different experimental conditions. The database contains more than 800 entries for the protein lysozyme, for example.

A polymer does not necessarily assume a unique three-dimensional structure in solution, which is equivalent to saying that its energy landscape (the value of the free energy for each possible arrangement of its atoms) does not necessarily have a single, global free-energy minimum. However, a protein is not just any polymer. It is a special polymer in that, in a given environment and physiological conditions (pH, temperature, ionic strength, etc.), it assumes one, and only one, specific three-dimensional structure. Some important implications and limitations of this statement are discussed in Problem 4.

Figure 1 shows the experimentally determined three-dimensional structure of a protein in which each atom is depicted as a sphere (see color insert after page 40). The protein is glycogen phosphorylase, one of the enzymes that allow us to survive without feeding continuously, even though our cells need

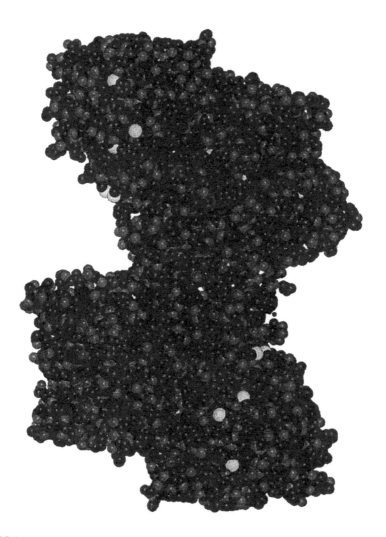

FIGURE 1

An all-atom representation of a protein structure determined by X-ray crystallography. This protein is an enzyme, glycogen phosphorylase from rabbit muscle, and its code in the Protein Data Bank is 1ABB. Atoms are colored according to a commonly used scheme: carbon is black, nitrogen is blue, oxygen is red, and sulfur is yellow.

a constant supply of sugars. The sugars that we consume are stored in our muscles in the form of glycogen, a macromolecule that contains up to 10,000 glucose molecules. The glycogen granule is clipped into glucose, and this chemical reaction is catalyzed by glycogen phosphorylase.

At first sight, the structure appears very complex, with no apparent regularities. Hopefully, though, by the end of this Introduction, the reader not only will be fascinated by the beauty of this molecule and impressed by its versatility, but also will have learned how to detect the underlying regularities as well as some general properties of protein structures.

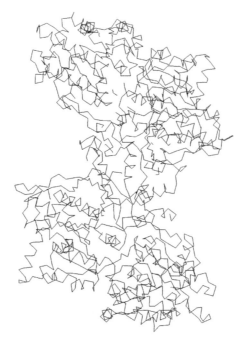

FIGURE 2
A different representation of the protein in Figure 1. This time, each amino acid is depicted as a vertex of a broken line connecting the amino acid chain.

The remainder of this Introduction is devoted mainly to the structure of proteins that spend their time in polar environment. Apolar proteins, which are embedded in biological membranes, and their structural properties are discussed in Problem 5.

Figure 2 shows the same protein as in Figure 1, but now, instead of every atom, only one atom per amino acid is shown as a vertex of a broken line that connects equivalent atoms in consecutive amino acids. The atom selected in the figure is called Cα and, in amino acids, it is linked to four different chemical groups: a carboxylic group, an amidic group, a hydrogen atom, and a variable chemical group (the side chain). The amino acids that occur in natural proteins number exactly 20 and differ by their side chains (as shown in Figure 3). The side chain can be a single hydrogen atom, as in the case of glycine, or can contain polar, neutral, and charged groups. Hydrogen atoms are usually not shown in protein-structure representations, because their positions are difficult to detect by X-ray crystallography.

Amino acids are linked to each other by a chemical bond, the peptide bond, between the carboxylic group of one amino acid and the amidic group of the adjacent amino acid. The chemical chain formed by the amidic group, the Cα, and the carboxylic group is called the main chain, or backbone, of the polypeptide. Different side chains protrude from the backbone, and their

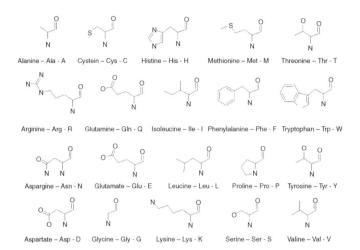

FIGURE 3
The 20 naturally occurring amino acids.

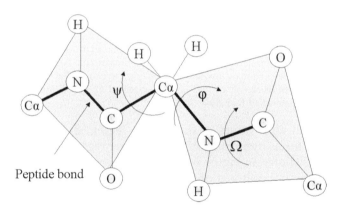

FIGURE 4
Amino acids linked by a peptide bond. The thicker lines form the "backbone" of a protein.

sequence defines the properties of the protein. The sequence of amino acids of a protein is called its primary structure.

The backbone of a polypeptide chain is quite flexible (as can be appreciated by looking at Figure 2). However, only the two angles ϕ and ψ (Figure 4) can assume several conformations in solution, the remaining angle, around the peptide bond, is planar. Furthermore, not all combinations of the values of ϕ and ψ are energetically favorable. Some are rarely observed, as is shown in Figure 5.

The combinations ϕ −60 and ψ −50 and ϕ −110 and ψ 130 are energetically favorable and observed very often. A consecutive stretch of residues with ϕ and ψ values in the first region, called the α_R region, assumes a helicoidal

a)   b)

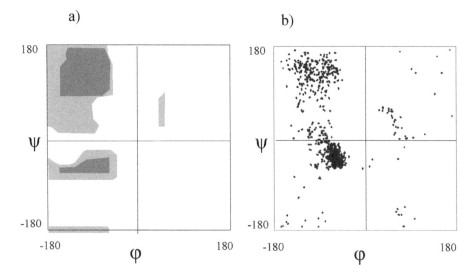

FIGURE 5

On the left is a representation of the results of energetic calculations of all possible pairs of φ and ψ angles in a dipeptide formed by alanine residues. Some combinations are energetically favorable (dark-gray areas) or allowed (light-gray areas), whereas others are unfavorable (white areas). This arrangement is reflected by the frequency at which combinations are observed in experimentally determined protein structures, as on the right side of the figure, where each point represents a φ and ψ pair observed in glycogen phosphorylase. The region with φψ angles around (60, 60) is rarely observed, and it is generally unfavorable because it brings the first carbon atom of the side chain too close to the carboxylic oxygen. The amino acid glycine does not contain a carbon in its side chain and is often observed in this conformation. The graphs shown in the figure are called Ramachandran plots.

shape. A stretch with values in the second region, the β region, becomes elongated and forms hydrogen bonds with other regions with the same local structure, as is shown in Figure 6 (see color insert after page 40).

Note that the polar atoms of the backbone (the carboxylic and amidic group) of both the α-helix and the β-sheet form hydrogen bonds with other main-chain atoms. This behavior is a result of the fact that, in the unfolded state and in an aqueous environment, the polar atoms would form hydrogen bonds with the surrounding water molecules. When the protein folds (i.e., when it assumes a compact shape), some of these atoms are shielded from the solvent and unable to form hydrogen bonds with it. This energy loss has to be compensated by the formation of hydrogen bonds within the protein chain.

A protein chain, in general, contains both hydrophobic and hydrophilic atoms. Exposure of the former to a polar solvent is energetically unfavorable, because a loss of entropy results (Figure 7). During folding, an energy gain is associated with the shielding of these groups from the solvent in addition to an energy gain through internal interactions (vanderWaals, charge–charge, and intrachain hydrogen bonds) established in the final structure. In proteins, these interactions are sufficient to compensate for the loss of entropy

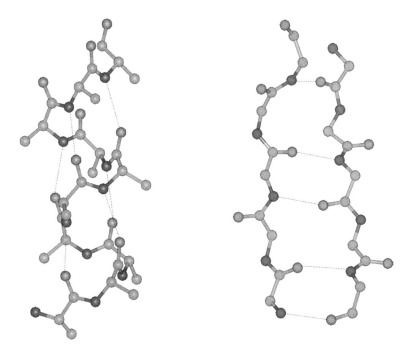

FIGURE 6
The backbone atoms of an α-helix and of two β-strands are depicted above. The strands, pairing via hydrogen bonds (dotted lines), form a β-sheet.

associated with folding (an unfolded chain has practically an infinite number of possible conformations and, therefore, a very high entropy), and a unique three-dimensional structure (called tertiary structure) can be achieved.

In a protein structure, polar amino acids (i.e., amino acids whose atoms can form hydrogen bonds with the water) are found more often at its surface, while the hydrophobic amino acids are mostly buried inside (Figure 8) (see color insert after page 40).

Most proteins contain regions in α and β conformations, collectively called regions of repetitive secondary structure, and connecting regions called loops. We can further modify our view of proteins by using cylinders and arrows to depict the secondary structure elements, as is shown in Figure 9 (see color insert after page 40). The latter representation of our protein shows its beautiful regularity. Two seemingly identical chains (a protein formed by more than one amino acid chain has a quaternary structure) are formed by two "lobes"; that is, structural regions that have more contacts between themselves than with other regions of the protein. We call these "lobes" domains. The two domains are not identical, but they show some topological similarity and can be described as three-layered structures, two external helical parts and a central β-sheet.

Cellular mechanisms can chemically modify a protein's primary structure after it has been synthesized. These modifications can be permanent or can

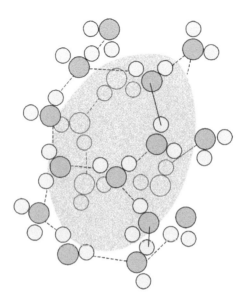

FIGURE 7
The hydrophobic effect. Polar molecules (water in the figure) form many energetically favored hydrogen bonds. When an hydrophobic molecule is present, they organize themselves around it in a more ordered way and therefore lose entropy.

vary according to the cellular state. For example, several proteins are glycosylated; that is, sugar chains are covalently linked to their amino acids. Another common protein modification is phosphorylation, which is often used for regulation, signal transduction, and cell cycle regulation. These modifications can affect the protein structure, often to a very considerable extent, and their presence and precise localization can depend on specific patterns of amino acids.

Glycogen phosphorylase is one such example. When a phosphate molecule is added to a serine amino acid (serine 14), shown in Figure 10 as a green sphere (see color insert after page 40), a shift occurs in the structural elements of the enzyme. This conformational change activates the protein. Phosphorylation of this enzyme is performed by other enzymes that monitor the concentration of sugar in the blood. The activity of the protein also has to increase when the energy levels of the cell are low. AMP (adenosine monophosphate) is a product of ATP (adenosine triphosphate) breakdown, an energetically favorable chemical reaction that provides energy to the cell. More AMP is created when energy levels are low and more sugar is needed. Binding of AMP to a site on glycogen phosphorylase causes similar structural changes as phosphorylation and activates the enzyme.

The Structure–Function Relationship in a Protein

The amino acid sequence of a protein contains amino acids selected for shaping its energy landscape that specify the unique three-dimensional native

FIGURE 8
A section of the structure of SH3, a small module found in many proteins, where it acts as an adapter to recruit other proteins. The green hydrophobic amino acids are more frequent in the inside than on the outside of the molecule.

structure and do not allow chains to fold into undesired conformations or not achieve a definite structure at all. The amino acid sequence also contains the specific residues necessary for the protein's function.

Without delving into the chemical details of the enzymatic activity of glycogen phosphorylase, we mention that it needs the close proximity of four amino acids (see the inset in Figure 10): two lysine residues at positions 568 and 574, one arginine at position 569, and one threonine at position 676. These amino acids are distant in the linear amino acid chain, but are brought together in the precise relative position needed for catalysis by the three-dimensional structure of the protein in its active form (i.e., when the protein is phosphorylated or when AMP is bound). The region of an enzyme where action takes place is called its active site. The structure of the active site in the inactive form of glycogen phosphorylase (i.e., the relative position of the catalytic residues) has a somewhat different structure than in its active form. The enzyme is, in fact, less efficient. Incidentally, the existence of two structures

FIGURE 9
The structure of glycogen phosphorylase once again. This time helices and strands are shown as cylinders and arrows.

for the enzyme does not contradict what we said about the uniqueness of the protein shape, because the two conformations are achieved with different ligands and, therefore, are not in the same environmental conditions.

Another example of the importance of the three-dimensional structure for function is illustrated in Figure 11 (see color insert after page 40). The reader should now see that the depicted protein is mostly formed by β-strands. It has two domains and a quaternary structure. It is an enzyme encoded by the virus responsible for hepatitis C. This virus enters the host cell and synthesizes a single, long amino acid chain that is later broken into smaller fragments, each of which encodes one of its functions. The enzyme shown in the figure breaks up the long polyprotein. It is a protease; that is, an enzyme that catalyzes the cleavage of peptide bonds.

We can now look at its active site. The atoms responsible for the catalytic activity belong to three amino acids: a serine, a histidine, and an aspartic acid. It also has a "pocket" ideally suited for accommodating the side chain of one specific amino acid, cysteine, and, in this way, it recognizes the precise location where cleavage should occur. In the figure is an enlarged view of the region involved in recognition and catalysis. The involved amino acids come from different parts of the amino acid chain and, once again, the three-dimensional structure of the protein allows them to be correctly positioned.

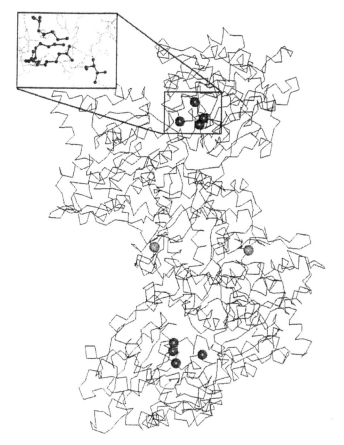

FIGURE 10
The active site of glycogen phosphorylase. The phosphorylation of serine 14, shown as a green ball, triggers a conformational change in the protein.

One way to inhibit the activity of this enzyme and, thereby, interfere with the function of the virus is to design a molecule that occupies and blocks the site where the enzyme binds the target amino acid (a cysteine). Detailed knowledge of the three-dimensional structure of the protein is fundamental to designing such a molecule.

Detection of the residues responsible for function (e.g., those that form an active site or an interaction surface) solely on the basis of a protein's amino acid sequence is practically impossible. These residues are no different from other amino acids. Only their specific positioning in the context of the final three-dimensional structure allows them to perform their function. Yet, the goal we are pursuing is detection of the sites important for activity and the understanding of how a function is performed, in the absence of an experimental three-dimensional structure.

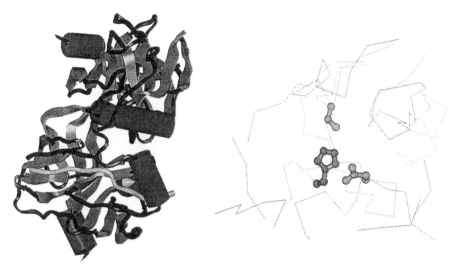

FIGURE 11
The structure of the protease of the hepatitis C virus (PDB code: 1NS3).

Suggested Reading

Berg, J.M., Stryer, L. and Tymoczko, J.L. *Biochemistry*, 5th ed., W. H. Freeman, New York, 2002.

Branden, C. and Tooze, J. *Introduction to Protein Structure*, 2nd ed., Garland Publishing, New York, 1999.

Creighton, T.E. *Proteins*, 2nd ed., W.H. Freeman, New York, 1993.

Nelson, D.L. and Cox, M.M. *Lehninger Principles of Biochemistry*, 4th ed., W. H. Freeman, New York, 2004.

Lesk, A.M. *Introduction to Protein Architecture—The Structural Biology of Proteins*, Oxford University Press, Oxford, 2000.

Voet, D. and Voet, J. *Biochemistry*, 3rd ed., Wiley, New York, 2004.

Drenth, J. *Principles of Protein X-ray Crystallography*, 2nd ed., Springer, New York, 1999.

Rhodes, G. *Crystallography Made Crystal Clear*, Academic Press, New York, 1999.

Wuthrich, K. *NMR of Proteins and Nucleic Acids*, Wiley-Interscience, New York, 1986.

Cavanagh, J., Arthur, W.J.F., Palmer, III, G., and Skelton, N.J. *Protein NMR Spectroscopy: Principles and Practice*, Academic Press, New York, 1995.

Bernstein, F.C., Koetzle, T.F., Williams, G.J., Meyer E.F.J., Brice, M.D., Rodgers, J.R., Kennard, O., Shimanouchi, T., and Tasumi, M. The Protein Data Bank: A computer-based archival file for macromolecular structures, *Eur. J. Biochem.* 80, 319–324, 1977.

Acknowledgments

Numerous people have generously offered me advice and support. I express my gratitude to all colleagues of the Department of Biochemical Sciences of the University of Rome "La Sapienza." In particular, I thank professors Maurizio Brunori, Francesca Cutruzzolà, and Carlo Travaglini-Allocatelli and doctors Veronica Morea, Romina Oliva, and Simonetta Soro. My most special thanks go to Dr. Domenico Cozzetto for his dedication and patience in critical reading of the manuscript.

Contents

Problem 1

Protein Sequence Alignment

Introduction to the Problem

The Evolution of Proteins

The sequences of amino acids in naturally occurring proteins have been selected by evolution for their favorable thermodynamic, kinetic, and functional properties. Variations in protein sequences are continuously generated via several molecular mechanisms. When variations that do not impair essential functions occur in germinal or replicating cells, they are transmitted to the progeny and generate diversity in the population, whereas variations that do impair essential functions disappear. Variations can be caused by substitution of one DNA base with another, replication of whole regions of a genome, and insertion and deletion of bases. In diploid organisms, which have two chromosome sets and, therefore, two copies of the genetic material, a mechanism known as crossing-over (i.e., the exchange of regions between two homologous chromosomes) can also introduce diversity.

Because of the degeneracy of the genetic code, a single base substitution in a protein-coding gene might or might not lead to an amino acid replacement or to the replacement of one amino acid with one of the translation–termination codons. Insertions and deletions might cause a frame shift in the gene, and this process can modify the downstream sequence of the encoded protein (because each amino acid is coded by three bases, and no spacing signal exists between coding triplets) or result in the addition or deletion of amino acids in the protein (Figure 12).

Individuals of a population can diverge sufficiently to give rise to different species. Speciation is an ill-defined concept in biology. Two individuals are operationally defined as belonging to different species if they cannot produce offspring when mating in the wild. This inability to reproduce is not strictly dependent upon genetic difference; environmental or physical factors could also account for inability to mate.

```
atggccctgtggatgcgcctcctgcccctgctggcgctgctggccctctggggacctgac
 M  A  L  W  M  R  L  L  P  L  L  A  L  L  A  L  W  G  P  D
ccagccgcagcctttgtgaaccaacacctgtgcggctcacacctggtggaagctctctac
 P  A  A  A  F  V  N  Q  H  L  C  G  S  H  L  V  E  A  L  Y
ctagtgtgcgggggaacgaggcttcttctacacacccaagacccgccgggaggcagaggac
 L  V  C  G  E  R  G  F  F  Y  T  P  K  T  R  R  E  A  E  D
ctgcaggtggggcaggtggagctgggcggggggccctggtgcaggcagcctgcagcccttg
 L  Q  V  G  Q  V  E  L  G  G  G  P  G  A  G  S  L  Q  P  L
gccctggaggggtccctgcagaagcgtggcattgtggaacaatgctgtaccagcatctgc
 A  L  E  G  S  L  Q  K  R  G  I  V  E  Q  C  C  T  S  I  C
tccctctaccagctggagaactactgcaactag
 S  L  Y  Q  L  E  N  Y  C  N  -
```

```
atggccctgtggatgcgcctcctgcccctgctggcgctgctggccctctggggacctgac
 M  A  L  W  M  R  L  L  P  L  L  A  L  L  A  L  W  G  P  D
ccagccgcAgacctttgtgaaccaacacctgtgcggctcacacctggtggaagctctcta
 P  A  A  D  L  C  E  P  T  P  V  R  L  T  P  G  G  S  S  L
cctagtgtgcgggggaacgaggcttcttctacacacccaagacccgccgggaggcagagga
 P  S  V  R  G  T  R  L  L  L  H  T  Q  D  P  P  G  G  R  G
cctgcaggtggggcaggtggagctgggcggggggccctggtgcaggcagcctgcagccctt
 P  A  G  G  A  G  G  A  G  R  G  P  W  C  R  Q  P  A  A  L
ggccctggaggggtccctgcagaagcgtggcattgtggaacaatgctgtaccagcatctg
 G  P  G  G  V  P  A  E  A  W  H  C  G  T  M  L  Y  Q  H  L
ctccctctaccagctggagaactactgcaactag
 L  P  L  P  A  G  E  L  L  Q  L
```

FIGURE 12

The first panel shows the nucleotide sequence and the corresponding translation in amino acids of human insulin, a protein that participates in the metabolism of fat and proteins. In the second panel, a base is inserted in the gene (an A shown in uppercase and bold). Note that this insertion affects the translation of the whole downstream region of the proteins.

If the modified protein allows survival of the progeny, the mutation is accepted in the population, and it can become the most frequent variant if it confers a selective advantage to the individual. This development implies that we can observe only those variants of a protein sequence compatible with the functionality of the organism and derived from a functional ancestor via a set of functional, or at least nondeleterious, variations.

In Problem 4, we discuss in which cases and to what extent this observation allows us to use the amino acid sequence of a protein to infer its three-dimensional native structure and how the latter can be used to infer function. Here, we state that this problem has no general solution, and, therefore, no standard route leads from sequence to structure to function.

However, if the function performed by a protein has to be conserved and function is brought about by specific amino acid residues and by their relative position in the three-dimensional structure, then residues responsible for function and structure must be conserved during evolution. This observation immediately suggests a strategy for detecting these residues.

Evolution-Based Inference of Protein Function

If we can identify an evolutionary relationship between two proteins from different species and highlight conserved amino acids (i.e., amino acids that could not be replaced in evolution without negatively affecting the protein's function), these amino acids are likely candidates for involvement in functional mechanisms. Clearly, we need to distinguish between amino acids that have been conserved because of their functional role and amino acids that have been preserved because of their structural role. In general, the constraints on the former are more stringent. Similar amino acids can more easily replace each other in a structural role than in a functional role; for example, catalysis requires specific atoms and, therefore, specific amino acids.

Furthermore, if one of the two proteins has been biochemically characterized, and its functional residues are known, residues corresponding to them (i.e., descending from the same amino acid of a parental ancestral protein) in the other protein are likely to perform the same role, if conserved.

This finding implies that we must determine how likely two proteins are to be evolutionarily related (homologous) (i.e., descending from a common ancestor protein) and determine the correspondence between the amino acids of the two proteins that most likely reflects their evolutionary history.

These two problems are known as "homology detection" and "protein sequence alignment," respectively, and can be formalized as follows:

1. Given two protein sequences Pa and Pb, calculate the probability that they are homologous (i.e., that a common ancestor protein has originated them via mutational events).

2. Given two protein sequences Pa and Pb, known to be homologous, identify all pairs of amino acids of the two proteins that derive from the same amino acid of the common ancestor.

In practice, the two problems are best reformulated as a single problem: given two protein sequences Pa and Pb, identify the correspondence between all pairs of their amino acids that maximizes the probability that they derive from the same amino acid of a common ancestor, and calculate the probability that such an ancestor exists.

In other words, we hypothesize that the two proteins are related and, therefore, that an evolutionary correspondence exists between their amino acid sequences (what we call their sequence alignment). We generally assume that the optimal correspondence is the one that requires the minimum number of mutational events or, equivalently, the one that maximizes the number of identical or similar amino acids. From this optimal alignment, we can calculate the probability that the observed similarity is likely to be statistically significant (i.e., unlikely to be observed by chance alone when two unrelated sequences are aligned).

If we were dealing with a merely statistical problem, we could simply compare the observed similarity with that expected for two random strings of the same length as our proteins and formed by a random combination of 20 characters. However, natural protein sequences are not random polymeric sequences, and, therefore, we must compare the similarity with that expected between two unrelated *protein* sequences, and these sequences are not randomly generated amino acid sequences.

Orthology and Paralogy

Imagine the following two scenarios: in scenario 1 (Figure 13a), gene *Ga* codes for protein *Pa* that performs the essential function *Fa*. After a speciation event, the protein evolves independently in each species. The sequence of the genes undergoes a series of mutation events that change the sequence into *Ga'* in one species and into *Ga''* in the other. The two species have two related proteins with sequence *Pa'* and *Pa''*, both compatible with the essential function *Fa* because only mutations that do not impair it are accepted and transmitted to the progeny. *Pa'* and *Pa''* might still share a sufficient sequence similarity to allow their evolutionary relationship to be detected.

In scenario 2 (Figure 13b), gene *Ga* codes for protein *Pa* that performs the essential function *Fa*. The region that contains *Ga* is duplicated, which generates an identical copy of the gene (*Gb*) that produces protein *Pb*. *Pb* is initially identical to *Pa*. After a speciation event, both species evolve inde-

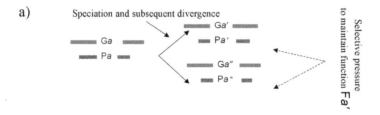

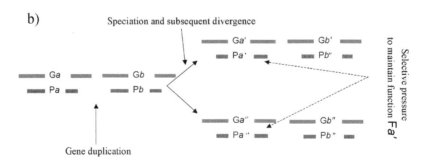

FIGURE 13
Evolution of paralogous and orthologous genes.

pendently, and *Pa* and *Pb* undergo a certain number of mutation events that generate *Pa'* and *Pb'* in one species and *Pa''* and *Pb''* in the other species. In each species, only one of the two proteins is subjected to the evolutionary pressure of preserving the essential function *Fa*. The other copy is free to evolve, and mutations that impair its function are accepted and transmitted to the progeny, provided the mutations are not deleterious to the organism.

The proteins not subjected to the evolutionary pressure might evolve a different function, not necessarily the same function and not necessarily related to the function *Fa*, but they can all still share a sufficient similarity between themselves and with the other two proteins that perform function *Fa* to allow their relationship to be detected.

Pa' and *Pa''* are called orthologous proteins; all other pairs of evolutionarily related proteins in the example are called paralogous proteins.

The issue should now be evident: the detection of an evolutionary relationship between two proteins does not guarantee that they share a common function because of the possibility of paralogous relationships. Therefore, we must distinguish which is the case in the problem at hand. If we are looking for the function of *Pa''* or *Pb''*, the discovery that they are evolutionarily related to *Pa'* might mislead us to conclude that they perform the function *Fa*.

Duplication, and subsequent divergence, is one of the biological mechanisms by which proteins develop new functions. Another route is the mixing and matching of domains. These compact protein substructures, assumed to fold independently, often have a functional role. Most proteins are built of several domains, and the function of a protein can be brought about by the cooperation of "subfunctions" performed by the domains. For example, one domain can recognize a specific region of the DNA and another domain can activate transcription after binding, or one domain can recognize the presence of a ligand and transmit this information to a second domain that, in turn, activates a different protein. The insertion of a functional domain into an existing protein can, therefore, change or modify the protein's biological function. An evolutionary relationship between two proteins limited, for example, to one domain does not guarantee that they have the same function.

Protein Families

As two protein sequences diverge, they accumulate changes, and the number of conserved amino acids between them might decrease their similarity to the level expected for two unrelated proteins. Nevertheless, detecting very distant homologous relationships is of paramount importance for at least three reasons.

First, detection of such relationships enlarges the number of proteins for which functional inference can be made. Second, detection of functionally important regions is made easier. If the divergence time has been very long, the strong evolutionary pressure to preserve functional residues becomes

```
Sequence 1        AL KTLNYDF DHLVEMESDAGL GNGGL GRLAACYLDSMATLAV
Sequence 2        VMK EFDL DLNEI I EQEPDPGL GNGGL GRLAACFLDSL ASLEV
Common residues    K       D        E  E  D  GLGNGGLGRLAAC  LDS  A   L   V


Sequence 1        AL KTL NYDF DHLVEMESDAGL GNGGL GRLAA CYL DSMAT LAV
Sequence 4        AYF SAEF GVHETL PI YS- GGL- - - - - GVLAGDHVKSA SDLNL
Common residues A                    S    GL        G  LA         S


Sequence 1        AL K TL NYDF DHLVEMESDAGL GNGGL GRLAACYL DSMATL AV
Sequence 2        VMK EFDL DLNEI I EQEPDPGL GNGGL GRLAACFL DSL ASL EV
Sequence 3        AL MDL GFK L EDLYDE ERDAGL GNGGL GRLAAC- MDSL ATCNF
Sequence 4        AY F SAE FGVHETL PI YS- - - - - - - GGLGVLAGDHV KSA SDL NL
Common residues                              GGLG  LA          S
```

FIGURE 14
The first panel shows a pairwise alignment between two evolutionarily related sequences. The two sequences are very similar and, therefore, easy to align. However, their similarity is such that most of their residues are identical, and, therefore, a determination of which are really important for function is difficult. The second panel shows the alignment of two distantly related sequences. This alignment is more useful in highlighting important residues but is more ambiguous. The third panel shows a multiple alignment between four sequences. This alignment is a better compromise because it is more reliable, and the conserved residues are easier to detect. The last part of the figure shows a useful graphical representation of a multiple-sequence alignment as a stack of symbols, one stack for each position in the sequence. The height of symbols within the stack indicates the relative frequency of each amino acid at that position.

more apparent against the background of the remaining positions that experienced a larger number of accepted mutational events, as illustrated by the example in Figure 14 (see color insert after page 40). Third, the detection of very distant relationships might reveal unexpected evolutionary links between organisms, which can increase understanding of how life developed.

One strategy for identifying distantly related proteins is based on the observation that homology is transitive: two proteins evolutionarily related to a third protein are evolutionarily related to each other. We can proceed by first identifying proteins likely to be homologous to our target sequence, then by identifying other sequences similar to the former, and so on. This process allows us to identify a larger number of proteins that share an evolutionary relationship (i.e., a larger family of proteins), as is shown by the simplified examples in Figure 14 and Figure 15 (see color insert after page 40).

We can align a whole family of proteins together; that is, we can obtain a correspondence table between the amino acids of all the proteins at the same

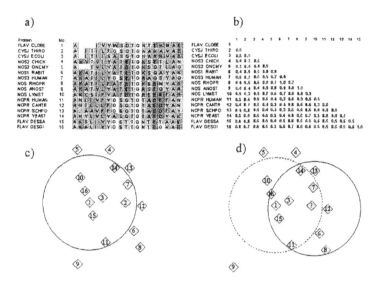

FIGURE 15

"Hopping" in sequence space. (a) A partial alignment of a protein sequence with a set of evolutionarily related proteins. (b) A matrix in which the percentage of different amino acids between each pair of sequences is charted. If we make the simplifying assumption that a difference lower than 0.4 is statistically significant, we can then conclude that sequences 2, 3, 7, 10, 11, 14, 15, and 16 are likely to be evolutionarily related to sequence 1. On the other hand, because homology is transitive, proteins evolutionarily related to sequence 2 are also related to sequence 1. We can highlight all sequences that are statistically more similar than expected to sequence 2 and, thereby, add sequences 6, 8, 12, and 13 to the family.

time. This procedure can highlight important properties of the family because each pair of sequences contains amino acid conserved for functional reasons and amino acids conserved by chance, but the latter are different for each of the pairs and, therefore, easier to identify.

Similarity Matrices

In the previous discussion, we assumed that two amino acids can be either identical or different, but this assumption is clearly a simplification. Pairs of amino acids can have more or less similar chemical properties. In two homologous proteins, a positively charged amino acid is more likely to be replaced by another positively charged amino acid than by a large hydrophobic residue, and this circumstance should be taken into account when one evaluates the probability that a sequence alignment corresponds to a true evolutionary relationship.

We must, therefore, estimate the probability that one amino acid is replaced by another during evolution. These values are empirically derived and reported in tables called similarity or substitution matrices. In these matrices, each row and each column corresponds to 1 of the 20 amino acids, and each cell contains a measure of the probability that the amino acids in the column and in the row can replace each other during evolution.

The method used to derive these probability values is straightforward: if we are given a set of aligned evolutionarily related sequences, we can calculate, for each pair of amino acids i and j, the frequency, $f_{i,j}$, with which the two amino acids are found in corresponding positions in the alignments (i.e., have replaced each other during the evolution of the protein family). On the other hand, the product $f_i f_j$ between the frequencies f_i and f_j, with which the amino acids i and j occur in the sequences, is an estimate of the probability that they are found in the same column by chance alone, given the composition of the sequences in the alignment. The ratio $f_{ij}/f_i f_j$ is an estimate of the likelihood that the amino acids i and j are substituted by each other during evolution. Similarity matrices usually report the $\log_2$ of these numbers. We need an initial alignment, or set of alignments, to derive the substitution values, and the alignments should be unambiguous.

The set of matrices proposed in 1978 by M.O. Dayhoff is based on the concept of PAM (point accepted mutation) and is called the PAM matrices. PAM is a measure of evolutionary distance between two proteins. An accepted point mutation is a single amino acid substitution that has been transmitted to the progeny (i.e., evolutionarily accepted). Two sequences are at 1 PAM distance if they can be converted into each other, assuming an average of 1 PAM every 100 amino acids.

The PAM1 matrix is constructed by deriving the substitution frequencies from alignments between pairs of proteins 1 PAM from each other, (i.e., very similar and, therefore, easy to align manually). The PAM2 matrix can then be obtained by multiplying the 1-PAM matrix by itself, the PAM3 by multiplying the PAM2 matrix by the PAM1 matrix, and so on, iteratively. The larger the number of the matrix, the more suitable it is for detecting more distant evolutionary relationships (Figure 16a).

The BLOSUM (blocks substitution matrix) matrices are derived by use of local alignments of very conserved regions in homologous proteins. They are also constructed as a series of matrices. A BLOSUM-N matrix is derived from alignments such that all sequences sharing more than N% identity with any other sequence in the alignment are averaged and represented as a single sequence. Unlike the PAM, here a larger number indicates that a matrix is more suitable for aligning closely related sequences (Figure 16b).

These matrices can be used to derive multiple-sequence alignments, which can, in turn, be used to refine the substitution frequencies between pairs of amino acids and derive new matrices. One example of this approach is represented by the Gonnet matrices, in which PAM250 is initially used to obtain the initial sequence alignments used to derive a new matrix that is subsequently used to realign the sequences. The new alignments are then used to derive another matrix, and so on, iteratively.

a)

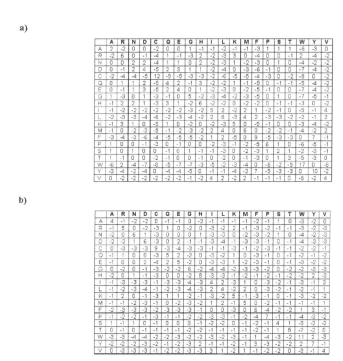

b)

FIGURE 16

(a) The PAM250 and (b) BLOSUM62 matrices.

Indel Penalties

Both the PAM and the BLOSUM matrices are derived from alignments of very similar protein sequences or of conserved regions of proteins and can be effectively used to statistically model single amino acid replacements. However, other evolutionary events are infrequent and are, therefore, rarely observed in very similar sequences. The events are the insertions and deletions of amino acids (also called indels or gaps).

As stated earlier, an alignment of two protein sequences is the correspondence between their amino acids that most likely reflects their evolutionary history. This history can include indel events, but these events should be penalized with respect to identities or single-residue substitutions because they are less frequently observed in homologous proteins. Proteins are very compact structures, and the insertion or deletion of one or more amino acids can be structurally destabilizing. Because substitution matrices are derived from ungapped alignments, indels are penalized by application of empirically derived parameters. The simplest penalization model assigns a constant penalty value to each inserted or deleted amino acid, but this model is not a very good model of biological reality, because insertions and deletions can be more easily accommodated in a protein structure if they occur near the solvent-exposed surface (i.e., in a limited set of positions.) Therefore, more realistic models of gap penalization treat the initiation of a gap differently from its continuation: increasing the length of a gap by one amino acid is penalized less than inserting one amino acid in a different position.

In the linear penalty model, each gap has a penalty w_o augmented by a penalty w_e lower than w_o for each increase in the length of the indel. Alternatively, each inserted and deleted amino acid in an indel can be assigned a penalty lower than the previous one; for example, $\log_e w_o$, in which w_o is the penalty assigned to the first inserted or deleted amino acid, and the index e runs from 2 for the adjacent inserted or deleted amino acid to l for the last amino acid in the gap. If $w_o = 5$, the penalty for inserting four amino acids is $5 + \log_2 5 + \log_3 5 + \log_4 5 = 5 + 2.32 + 1.46 + 1.16 = 9.95$. Compare this result with the value $5 \times 4 = 20$ that would be obtained by application of a linear penalty scheme.

Local versus Global Alignment

The alignment between two protein sequences is global when it is aimed at finding the optimal correspondence between all amino acids of both sequences and local when it attempts to find local regions of similarity between the two sequences. The latter is biologically relevant because it might allow us to detect evolutionarily related domains present in proteins whose remaining sequences have no evolutionary relationship or allow us to highlight regions that contain functional units subject to stronger evolutionary pressures. Local alignments are also useful for the detection of proteins homologous to a target protein in a large data set of unrelated proteins.

How Do We Align Sequences?

Global Alignment of Two Protein Sequences: The Needleman and Wunsch Algorithm

An alignment of two protein sequences is a correspondence between the amino acids or appropriately inserted gaps of the first sequence and amino acids or gaps of the second sequence. Clearly, a gap in the first sequence cannot correspond to a gap in the second sequence, and the alignment can be seen as a matrix with two rows, one corresponding to each sequence, and each column corresponds to a pair of aligned amino acids or to an amino acid and a gap.

We use a maximum-parsimony approach and assume that the best alignment (i.e., the one that best reflects the evolutionary relationship between the two protein sequences) requires the minimum number of substitutions, insertions, and deletions. If we use our substitution matrices to assign a score to each pair of amino acids (a score that reflects the probability that the amino acids replaced each other during evolution) and a penalty value for indels, we are looking for the alignment for which the sum of the scores of each pair of aligned amino acids, diminished by the penalty values for the indels, is maximum.

Thus, the problem of aligning two protein sequences is reduced to the problem of finding the alignment between the two strings that represent their sequences such that the global score is maximum, *given a score function and penalty values for indels*.

The optimal alignment between two strings can be exactly computed, but the optimal alignment is biologically correct only insofar as the score function and the indel penalty values are biologically reasonable.

The Needleman and Wunsch algorithm finds the optimal alignment between two sequences by calculating the optimal alignment between subsequences of increasing length. It is based on two assumptions:

- Mutations in different sites of a sequence occur independently.
- The length of a gap does not depend on the elements aligned to the gap.

Both hypotheses are approximations of biological reality because different positions in sequences are subject to different evolutionary pressure, and not all sequences of amino acids can be accommodated in inserted regions. These sequences are usually solvent exposed and, therefore, are more likely composed of hydrophilic or flexible amino acids.

However, under these assumptions and given a scoring matrix and a gap penalty scheme, the alignment problem can be exactly solved by dynamic programming methods.

Let us write the two sequences in the first row and the first column of a matrix (Figure 17). Each cell corresponds to the alignment of the amino acid in the row with that in the column and contains a score reflecting the similarity between the two amino acids. Our alignment problem can now be reformulated as follows: what is the set of cells (that is, pairs of aligned amino acids) that we should use to go from the upper left corner to the lower right corner so that we "collect" the larger score?

We must build a new matrix (the so-called cumulative matrix) in which each cell contains the maximum score achievable by any alignment that ends in that cell. This score is not difficult to compute.

If we know the maximum score that can be achieved by any alignment that ends in the cell $\{s_{i-1}, t_{j-1}\}$, $\{s_{i-1}, t_j\}$, and $\{s_i, t_{j-1}\}$, then the maximum score $F(i,j)$ that can be achieved by any alignment that includes the pair $\{s_i, t_j\}$ can be easily calculated. Only three "moves" allow extension of the alignment to the i,j position: aligning s_i and t_j, aligning s_i with a gap, and aligning t_j with a gap. We select the path that provides the maximum score:

$$F(i,j) = \max \begin{bmatrix} F(i-1,j-1) + \sigma(s_i, t_j) \\ F(i-1,j) - w_i \\ F(i,j-1) - w_j \end{bmatrix} \qquad (1)$$

where (s_i, t_j) is the score value for the pair of amino acids x_i and y_j, and w_i and w_j are the appropriate penalties for inserting a single amino acid gap in the i and j positions, respectively.

We must now compute $F(0,0)$, $F(1,0)$, $F(0,1)$. $F(0,0)$ is equal to zero, as it does not correspond to any pair of aligned residues. $F(1,0)$, $F(0,1)$ are the maximum scores that can be achieved by aligning the first amino acid of each sequence with a gap (Figure 17), and, therefore, we can set their values to $-w_i$ and $-w_j$ or to 0, if we do not want to penalize gaps at the beginning of either sequence. Similar considerations allow us to fill the complete first row and first column. They correspond to further insertions in one of the two sequences, so we can either assign an extra indel penalty for each of them or set them to 0. Now we can completely fill the matrix, and the maximum score of the global alignment is, by definition, the one reported in the lower right cell (n,m). We can achieve the maximum score if we include in the alignment the (n,m) pair and one of the pairs $(n-1,m)$, $(n,m-1)$, or $(n-1,m-1)$, namely, the one that we used to obtain the maximum value in (n,m) that we recorded during the matrix construction procedure. This step can be iterated until a cell of the first column or of the first row is reached, which generates a path though the matrix including the cells that contribute to the alignment with maximum score.

Local Alignment of Two Protein Sequences: The Smith and Waterman Algorithm

A modification of the Needleman and Wunsch algorithm can be used to obtain local sequence alignments; that is, alignments that do not necessarily include the whole length of the two sequences. The difference consists essentially of not allowing negative scores, because a good local alignment is unlikely to include gaps or rarely observed substitutions. From the algorithmic point of view, this assumption implies that equation (1) is replaced by

$$F(i,j) = \max \begin{bmatrix} 0 \\ F(i-1,j-1) - \sigma(s_i, t_j) \\ F(i-1,j) - w_i \\ F(i,j-1) - w_j \end{bmatrix} \qquad (2)$$

We assign a value of 0 to $F(i,0)$ and $F(j,0)$ for each i and j and construct the matrix as in the previous case. The reconstruction of the alignment now starts from the maximum value in the matrix and proceeds with the same strategy as in global alignments, but it stops when a score of 0 is encountered.

Multiple-Sequence Alignments

A multiple-sequence alignment (MSA) between N protein sequences is again a matrix, this time with N rows such that each row contains a protein sequence, possibly with gaps inserted. Clearly, each column should contain at least one element that is not a gap.

As for pairwise alignment, the problem is to find the MSA that maximizes a predefined score. Unfortunately, the extension of the scoring scheme and of the algorithms described for two sequences to multiple-sequence alignments presents a number of problems.

One possible scoring scheme, called SP-score (sum of pairs score), is calculated by addition of all possible pairwise scores for each column and then summation of the scores for all columns. The underlying assumptions for the SP-score is that both the columns and the rows of the alignment matrix are statistically independent. This assumption is clearly not valid, because the protein sequences in the alignment are supposedly evolutionarily related.

Another problem of SP-scores can be better illustrated by the following example. Let us calculate an SP-score for the alignment in Figure 18a. For simplicity, we assign a score of 1 to identical amino acids and 0 for all other amino acids. Each column has three amino acids. In the first column, they are all different, and the score is 0. In the second and third columns, all three pairs are formed by identical amino acids, and the score is $3 \times 1 = 3$. The fourth column has again a score of 0. The last column contains one match (*F-F*) and two mismatches (*F-Y*), and, therefore, the score is $1 \times 1 + 2 \times 0 = 1$.

The alignment in Figure 18b is biologically less convincing than the alignment in Figure 18a, because the third position contains one mismatch, and the SP-score is indeed lower. The ratio between the score of the two alignments is $7/6 = 1.17$.

Let us now assume that we add one more sequence to both alignments as shown in Figure 18c and Figure 18d. From a biological point of view, the addition of a fourth sequence where again a C appears in the third position makes the first of the two alignments even more convincing, whereas it decreases the relative likelihood of the second. The presence of the G (in bold in Figure 18) in an otherwise very conserved position is more unlikely here than in the alignment shown in Figure 18a and Figure 18b. We would like our scoring system to recognize that the addition of the fourth sequence shifts our confidence toward the first alignment, yet the ratio of the scores of the two alignments is now 1.08 and erroneously points to a smaller difference between the quality of the two alignments shown in Figure 18c and Figure 18d with respect to those shown in Figure 18a and Figure 18b. This problem and the fact that there is no statistical justification for its usage notwithstanding the SP-score is often used because of its simplicity.

Let us assume that we need to align two sequences:

ACFFTGHILPRG and ADYTGHLMPKA

We first build a substitution matrix: the two sequences are in the first row and in the first column. In each cell there is the score for the pair of amino acids corresponding to the column and the row derived from the PAM250 matrix shown in Table 2:

	A	C	F	F	T	G	H	I	L	P	R	G
A	2	-2	-3	-3	1	1	-1	-1	-2	1	-2	1
D	0	-5	-6	-6	0	1	1	-2	-4	-1	-1	1
Y	-3	0	7	7	-3	-5	0	-1	-1	-5	-4	-5
T	1	-2	-3	-3	3	0	-1	0	-2	0	-1	0
G	1	-3	-5	-5	0	5	-2	-3	-4	0	-3	5
H	-1	-3	-2	-2	-1	-2	6	-2	-2	0	2	-2
L	-2	-6	2	2	-2	-4	-2	2	6	-3	-3	-4
M	-1	-5	0	0	-1	-3	-2	2	4	-2	0	-3
P	1	-3	-5	-5	0	0	0	-2	-3	6	0	0
K	-1	-5	-5	-5	0	-2	0	-2	-3	-1	3	-2
A	2	-2	-3	-3	1	1	-1	-1	-2	1	-2	1

Let's decide that we select a constant penalty value of 5 for indels. Now we need to build the cumulative matrix. We will set the penalty for inserting at the beginning of each sequence to zero. Therefore,

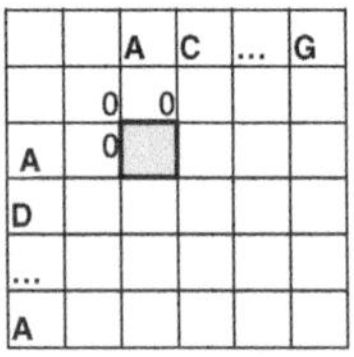

		A	C	...	G
	0	0			
A	0				
D					
...					
A					

We need to fill the shaded cell, and we can get there either from the cell on top of it, from the one at its left or from the cell diagonally up. If we came from the cell on the top, we would effectively insert one residue: the A of the horizontal sequence would not correspond to any amino acid of the vertical sequence. In this case the score in the cell should be – 5, i.e. 0 (the content of the cell we are coming from) minus the penalty for an indel, which is 5. The second case is similar (we would be inserting in the other sequence) and the score would again be – 5. In the third case there would be no insertion and the score would be 0 (the content of the cell we are coming from) plus 2 (the score of the A, A pair) =2. The maximum value between (-5, -5, 2) is 2, therefore we write 2 in the cell and remember that we obtained it using the cell diagonally up:

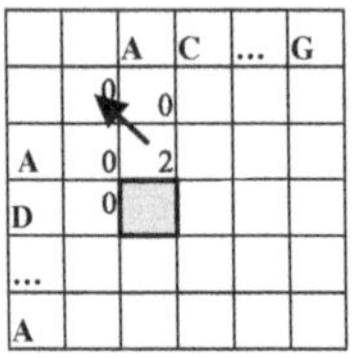

		A	C	...	G
	0	0			
A	0	2			
D	0				
...					
A					

FIGURE 17
The Needleman and Wunsch algorithm for pairwise sequence alignment. *(continues)*

We can now fill another cell, shaded in the scheme. If we came from the cell above it, the score would be 2 (the starting value) −5 (the insertion penalty) = −3. If we came from the cell on the left, we would have a score of 0 −5 = −5, for the diagonal cell the score would be 0 +0 = 0. Therefore:

And so on, until we complete the cumulative matrix:

	A	C	F	F	T	G	H	I	L	P	R	G
0	**0**	**0**	0	0	0	0	0	0	0	0	0	0
A	0	**2**	-2	**-3**	-3	1	1	-1	-1	-2	1	-2	1
D	0	0	-3	**-8**	-8	-3	2	2	-3	-5	-3	0	-1
Y	0	-3	0	**4**	-1	-6	-3	2	1	-4	-8	-5	-5
T	0	1	-4	-1	1	**2**	-3	-3	2	-1	-4	-9	-5
G	0	1	-2	-6	-4	1	**7**	2	-3	-2	-1	-6	-4
H	0	-1	-2	-4	-8	-4	-1	**13**	0	-5	-2	1	-4
L	0	-2	7	0	-2	-7	-6	8	**15**	10	5	0	-3
M	0	-1	2	7	0	-3	-8	3	10	**19**	14	9	4
P	0	1	-3	2	2	0	-3	-2	5	14	**25**	20	15
K	0	-1	-4	-3	-3	2	-2	-3	0	9	20	**28**	23
A	0	2	-3	-7	-6	-2	3	-2	-4	4	15	23	**29**

The optimal alignments, given our substitution matrix and our indel penalty scheme, are:

```
A D - Y T G H L M P K A
A C F F T G H I L P R G


A D Y - T G H L M P K A
A C F F T G H I L P R G


- A D Y T G H L M P K A
A C F F T G H I L P R G


- - A D Y T G H L M P K A
A C F - F T G H I L P R G
```

FIGURE 17 (CONTINUED)

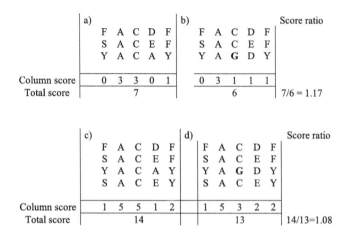

FIGURE 18

Illustration of the problems connected with the SP-score in multiple sequence alignments.

Several alternative scoring schemes have been proposed, but they all assume that each sequence is unrelated to any other sequence in the alignment; this assumption is not valid for sequences that belong to an evolutionary family.

Another nontrivial problem is connected with multiple-sequence alignments. A simple calculation shows that the extension of the Needleman and Wunsch algorithm to N sequences is technically unfeasible.

If we want to extend the algorithm to a multiple-sequence alignment of N sequences of length L, each cell has $2^{(n-1)}$ neighboring cells, and the total number of cells is L^N. Therefore, the calculation of the optimal alignment requires a number of steps of the order of $L^N \times 2^N$, a number that becomes quickly untreatable. An alignment of 50 sequences of length 150 amino acids requires more than 10^{150} operations.

Interesting ideas have been presented to solve this problem. For example, one can calculate the minimum of the alignment score for each pairwise alignment and assume that this score is a lower bound for the overall score. When we use our algorithm, we can reduce the number of cells that we have to consider by discarding those that would make our score lower than the lower bound. Even so, the problem can be solved in a reasonable time only if the number of sequences is not more than a few, and their length is only 100 to 200 amino acids.

Another solution to the problem is the use of a progressive method; that is, first align two sequences, then align a third sequence to the first alignment, then align a fourth to the alignment, and so on, iteratively. To align a sequence to an alignment, we can align the new sequence to each sequence of the alignment and select the highest scoring pair. We can align two alignments by calculating each possible pairwise score between sequences of the first alignment and sequences of the second, again using the highest scoring pair to build the new alignment.

We can also use a simple modification of the pairwise alignment algorithm. Instead of writing a single sequence in the first row, we write a previously calculated pairwise alignment. The score of each cell in the substitution matrix is the average score of each amino acid of the alignment with the corresponding amino acid of the vertical sequence:

	A	C	F	F	T	G	H	I	L	P	R	G
	A	D	Y	F	S	G	S	V	M	P	K	A
A	2.0	−1.0	−3.0	−3.0	1.0	1.0	0.0	−0.5	−1.5	1.0	−1.5[a]	1.5
D	0.0	−0.5	−5.0	−6.0	0.0	1.0	0.5	−2.0	−3.5	−1.0	−0.5	0.5
Y	−3.0	−2.0	8.5	7.0	−3.0	−5.0	−1.5	−1.5	−1.5	−5.0	−4.0	−4.0
T	1.0	−1.0	−3.0	−3.0	2.0	0.0	0.0	0.0	−1.5	0.0	−0.5	0.5
G	1.0	−1.0	−5.0	−5.0	0.5	5.0	−0.5	−2.0	−3.5	0.0	−2.5	3.0
H	−1.0	−1.0	−1.0	−2.0	−1.0	−2.0	2.5	−2.0	−2.0	0.0	1.0	−1.5
L	−2.0	−5.0	0.5	2.0	−2.5	−4.0	−2.5	2.0	5.0	−3.0	−3.0	−3.0
M	−1.0	−4.0	−1.0	0.0	−1.5	−3.0	−2.0	2.0	5.0	−2.0	0.0	−2.0
P	1.0	−2.0	−5.0	−5.0	0.5	0.0	0.5	−1.5	−2.5	6.0	−0.5	0.5
K	−1.0	−2.5	−4.5	−5.0	0.0	−2.0	0.0	−2.0	−1.5	−1.0	4.0	−1.5
A	2.0	−1.0	−3.0	−3.0	1.0	1.0	0.0	−0.5	−1.5	1.0	−1.5	1.5

[a] $= \frac{1}{2}[score\ (R.A) + score\ (K.A)] = \frac{1}{2}(-2-1) = -1.5$

We must now decide the order in which to align the sequences. In other words, which pair do we align first, and in which order do we select the subsequent sequences? Aligning the most similar pair of sequences first seems sensible because their correct alignment is likely to be easier and might, therefore, aid in the more difficult subsequent alignments.

The strategy is as follows:

1. Calculate similarity values for each pair of sequences.
2. Select the pair with highest similarity and proceed to align them.
3. Recalculate the similarity between the aligned pair and each of the other sequences.
4. Repeat steps 2 and 3 until all sequences have been aligned.

The result of the procedure can be visually represented as a binary tree, in which each node is a sequence and each edge is proportional to the "distance" between the nodes. The distance is inversely related to the similarity between two sequences, between one sequence and one alignment, or between two alignments. The tree is usually called the guide tree of the alignment and approximates the evolutionary relationship between the sequences.

Phylogenetic trees (i.e., trees representing true evolutionary relationships) are rooted; they include one node that represents the original ancestor of all sequences. Several algorithms are available for constructing trees, provided

In the matrix, S1, S2, S3, S4 and S5 represent five sequences and the numbers in each cell are the "distance" between the sequences of the row and the column. This can be some function of the similarity or identity value between the sequences. We select the two "closest" sequences and draw them as shown on the left in such a way that the vertical line is half of their distance.

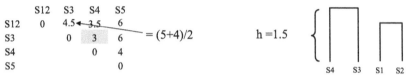

	S1	S2	S3	S4	S5
S1	0	2	5	4	7
S2		0	4	3	5
S3			0	3	6
S4				0	4
S5					0

We now calculate the distance of this group from each other sequence as the average of their distances from S1 and S2:

	S12	S3	S4	S5
S12	0	4.5	3.5	6
S3		0	3	6
S4			0	4
S5				0

$= (5+4)/2$

$h = 1.5$

The closest pair is now S3 and S4:

	S12	S34	S5
S12	0	4	6
S34		0	5
S5			0

And finally s12 and S34:

	S1234	S5
S1234	0	5.5
S5		0

$h = 2.75$

FIGURE 19
The UPMGA (unweighted pair-group average) method for building a tree.

that metrics between sequences have been defined. Some of the algorithms provide hypotheses about the location of the roots, while others do not and can be used to construct unrooted trees. The latter, although unable to give information about the evolutionary events that have generated the family of homologous proteins, are still very useful for simplifying the problem of multiple-sequence alignments.

Several methods exist for reconstructing the most likely evolutionary tree that has generated the observed differences between a given set of sequences. The methods take into account several factors, such as the possibility that different branches of the tree have different evolutionary rates. For the purpose of multiple-sequence alignment, we do not need a very accurate evolutionary reconstruction, and we can use less sophisticated methods, such as the one described in Figure 19.

The multiple alignment protocol originally devised by Feng and Doolittle consists of the following two steps:

1. Calculate the distances between each pair of sequences and use them to construct an approximate tree that will only be used to decide the order of alignments and, therefore, does not need to be especially accurate.

2. Align the pair of sequences, the pair of alignments, or one sequence and one alignment, starting from the child nodes of the tree, so that the most similar sequences are aligned first and the most dissimilar ones are aligned last.

In the Feng and Doolittle implementation of this procedure, gaps are penalized the first time they appear in an alignment but not when they are inserted in the same position in the subsequent alignment step (once a gap, always a gap). This practice is biologically sensible because all proteins in the tree are assumed to be homologous, which implies that their three-dimensional structures are topologically similar. Therefore, the position of a gap is structurally equivalent in all proteins, and if allowed in one pair, it also falls into an allowed region in the others.

Another reconstruction approach is to use the score of a column of the alignment (e.g., the SP-score) as the score in the alignment procedure. We only need to define the score for a gap-to-gap alignment (e.g., 0).

The widely used ClustalW method is essentially based on this last algorithm, with some clever additions. The alignment of each pair of sequences or alignments is built by use of a matrix appropriate for their evolutionary distance. Gap penalties depend on the amino acids observed in the column, so the presence of many hydrophilic or flexible residues in a column lowers the gap penalty in that position. Furthermore, the gap penalty is increased for columns that do not contain gaps, if gaps are present nearby in the alignment. Finally, the guide tree can be adjusted at the alignment stage on the basis of the scores of the alignments.

One problem with iterative multiple-sequence alignment is that the addition of a new sequence cannot modify the preexisting alignment. One solution, proposed by Barton and Sternberg, consists of taking out one sequence at a time and realigning it to the multiple alignment.

Profiles

Given a multiple-sequence alignment, we can derive the probability that each given amino acid is found in one of the aligned positions. We simply count the number of times each of the 20 amino acids appears in each column and divide the number of appearances by the number of aligned sequences. If the number of sequences is sufficiently high, these frequency values approximate the probability of finding any given amino acid in any position of the alignment.

When a new sequence is available, we can align it to the multiple alignment, calculate the probability that each of its amino acids is found in each of the columns, and multiply these values to calculate the probability that the sequence "fits" the profile.

Multiplying probabilities creates problems because they are always lower than 1 and their product quickly becomes a very low number. The best way to solve the problem is to calculate the logarithm of the values in the column. Because the logarithm of a product is the sum of the logarithms of its factors, our calculation of the "fitness" of the new sequence only requires additions.

One drawback of using logarithms is that if one amino acid never appears in a column its frequency is 0, and we face a problem because the logarithm of 0 is infinite. The most-used method of solving the problem is to add 1 to each of the frequency values (method of the pseudocounts).

Hidden Markov Models

The classical example used for explaining Hidden Markov Models is that of the "occasionally dishonest casino": in a game in which a player can bet on the rolling of a die, the player wins if the die lands on every side but the "6." If the die is fair, the player has 1/6 probability of losing and 5/6 probability of winning. A dishonest croupier could use a die that has a higher probability of landing on a "6," (e.g., 50%). To avoid being caught, the croupier can switch from a fair die to a loaded die with a certain frequency. For example, he can change the die from fair to loaded after 20 rolls and from loaded to fair after 10 rolls. This process is represented graphically in Figure 20.

Why is this model called a Hidden Markov Model? It is a Markov model because the state of the system is only influenced by the previous state, and it is hidden because, if we now see a series of rolls (e.g., 1, 3, 2, 3, 6, 6, 6, 4, 3, 2), we cannot know which states (fair, loaded, or a combination of both) produced the result.

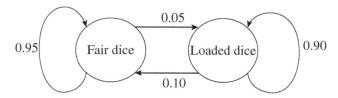

FIGURE 20
A Hidden Markov Model for the occasionally dishonest casino.

We can ask three questions about Hidden Markov Models:

1. Given a series of rolls, what are the probability values of the state transitions (i.e., how can the numbers associated with each of the arrows in Figure 20 be deduced)?

2. Given a series of rolls, what is the probability that the results have been generated by a scheme or model such as the one in Figure 20?

3. Given a series of rolls, what is the most likely sequence of states that generated it? In other words, which series of events (switching from the loaded to the fair dice and vice versa) most likely generated the observed results?

If we apply the scheme to a biological sequence alignment, we can envisage an alignment as a Hidden Markov Model in which each column is a roll. The outcome of the roll can be a match state (i.e., an amino acid is aligned to the next column), a delete state (i.e., the column is skipped), or an insert state (i.e., the amino acid does not correspond to any column). Our three questions can now be reformulated as follows:

1. Given a set of homologous amino acid sequences, which are the transition probability values (from match to insert, to delete and vice versa) that best describe the family (i.e., which is their Hidden Markov Model)?

2. Given an amino acid sequence, what is the probability that the sequence has been generated by a Hidden Markov Model of a protein family (i.e., what is the probability that the new sequence belongs to the previously aligned family)?

3. Given an amino acid sequence, what is the most likely sequence of matches, insertions, and deletions that makes the sequence fit to the Hidden Markov Model of a protein family? In other words, what is the most likely alignment of the new sequence to the existing model?

The best way to understand a Hidden Markov Model is to build one manually, as demonstrated below. The model is based on the alignment shown in Figure 15. The first eight columns of the alignment are shown in the matrix:

Position	1	2	3	4	5	6	7	8	...
Seq1	A	—	—	—	I	V	Y	W	
Seq2	A	—	I	T	I	L	F	G	
Seq3	A	—	I	T	I	I	S	A	
Seq4	A	K	V	T	V	I	Y	A	
Seq5	A	—	—	T	V	L	Y	A	
Seq6	A	K	A	T	I	L	Y	A	
Seq7	A	K	A	T	I	L	Y	G	
Seq8	A	K	A	T	I	L	F	A	
Seq9	A	K	A	T	V	L	Y	A	
Seq10	A	K	C	S	I	L	F	A	
Seq11	A	N	I	I	V	F	Y	G	
Seq12	A	N	T	L	L	L	F	G	
Seq13	A	—	A	A	V	F	F	G	
Seq14	A	N	Y	L	V	L	Y	A	
Seq15	A	K	S	L	I	V	Y	G	
Seq16	A	K	A	L	I	V	Y	G	

The first step is to count the frequency of each amino acid in each alignment position. Because some amino acids might be absent in some positions, we must add 1 to each of the values (pseudocounts):

Position	1	2	3	4	5	6	7	8	...
A	17	1	7	2	1	1	1	9	
C	1	1	2	1	1	1	1	1	
D	1	1	1	1	1	1	1	1	
E	1	1	1	1	1	1	1	1	
F	1	1	1	1	1	3	6	1	
G	1	1	1	1	1	1	1	8	
H	1	1	1	1	1	1	1	1	
I	1	1	4	2	10	3	1	1	
K	1	9	1	1	1	1	1	1	
L	1	1	1	5	2	10	1	1	
M	1	1	1	1	1	1	1	1	
N	1	4	1	1	1	1	1	1	
P	1	1	1	1	1	1	1	1	
Q	1	1	1	1	1	1	1	1	
R	1	1	1	1	1	1	1	1	
S	1	1	2	2	1	1	2	1	
T	1	1	2	9	1	1	1	1	
V	1	1	2	1	7	4	1	1	
Y	1	1	2	1	1	1	11	1	
W	1	1	1	1	1	1	1	2	

By calculating the frequencies, we obtain

Position	1	2	3	4	5	6	7	8	...
A	0.47	0.03	0.21	0.06	0.03	0.03	0.03	0.25	
C	0.03	0.03	0.06	0.03	0.03	0.03	0.03	0.03	
D	0.03	0.03	0.03	0.03	0.03	0.03	0.03	0.03	
E	0.03	0.03	0.03	0.03	0.03	0.03	0.03	0.03	
F	0.03	0.03	0.03	0.03	0.03	0.08	0.17	0.03	
G	0.03	0.03	0.03	0.03	0.03	0.03	0.03	0.22	
H	0.03	0.03	0.03	0.03	0.03	0.03	0.03	0.03	
I	0.03	0.03	0.12	0.06	0.28	0.08	0.03	0.03	
K	0.03	0.29	0.03	0.03	0.03	0.03	0.03	0.03	
L	0.03	0.03	0.03	0.14	0.06	0.28	0.03	0.03	
M	0.03	0.03	0.03	0.03	0.03	0.03	0.03	0.03	
N	0.03	0.13	0.03	0.03	0.03	0.03	0.03	0.03	
P	0.03	0.03	0.03	0.03	0.03	0.03	0.03	0.03	
Q	0.03	0.03	0.03	0.03	0.03	0.03	0.03	0.03	
R	0.03	0.03	0.03	0.03	0.03	0.03	0.03	0.03	
S	0.03	0.03	0.06	0.06	0.03	0.03	0.06	0.03	
T	0.03	0.03	0.06	0.26	0.03	0.03	0.03	0.03	
V	0.03	0.03	0.06	0.03	0.19	0.11	0.03	0.03	
Y	0.03	0.03	0.06	0.03	0.03	0.03	0.31	0.03	
W	0.03	0.03	0.03	0.03	0.03	0.03	0.03	0.06	

Now, we must count the number of transitions between two match states, a match and an insert, a match and a delete, and so on, on the assumption that the begin state is a match.

We must always remember to add the pseudocounts:

Position	1	2	3	4	5	6	7	8	...
Match–match	17	12	12	15	16	17	17	17	
Match–delete	1	6	1	1	1	1	1	1	
Match–insert	1	1	1	1	1	1	1	1	
Insert–match	1	1	1	1	1	1	1	1	
Insert–delete	1	1	1	1	1	1	1	1	
Insert–insert	1	1	1	1	1	1	1	1	
Delete–match	1	1	4	2	2	1	1	1	
Delete–delete	1	1	3	2	1	1	1	1	
Delete–insert	1	1	1	1	1	1	1	1	

Finally, we must convert the counts into frequencies:

Position	1	2	3	4	5	6	7	8	
Match–match	0.68	0.48	0.50	0.6	0.64	0.68	0.68	0.68	
Match–delete	0.04	0.24	0.04	0.04	0.04	0.04	0.04	0.04	
Match–insert	0.04	0.04	0.04	0.04	0.04	0.04	0.04	0.04	
Insert–match	0.04	0.04	0.04	0.04	0.04	0.04	0.04	0.04	
Insert–delete	0.04	0.04	0.04	0.04	0.04	0.04	0.04	0.04	
Insert–insert	0.04	0.04	0.04	0.04	0.04	0.04	0.04	0.04	
Delete–match	0.04	0.04	0.20	0.08	0.08	0.04	0.04	0.04	
Delete–delete	0.04	0.04	0.10	0.08	0.04	0.04	0.04	0.04	
Delete–insert	0.04	0.04	0.04	0.04	0.04	0.04	0.04	0.04	

We can now build the graphical representation of our Hidden Markov Model, in which the delete state is represented by a rhomboid, and the insert is represented by a circle, as shown in Figure 21. For clarity, only some of the probability values are indicated. The meaning of the Hidden Markov Model should now be apparent. For example, starting from the BEGIN state, we have a 68% probability of matching the first residue, and, if a match is chosen, the table can tell us the probability of aligning each of the 20 amino acids. After the first match, the probability of a deletion is 24%, and, if we indeed delete a residue, we have a probability of 10% of deleting another residue and a probability of 20% of matching it instead to a column of the alignment. We cannot know in advance which route will be taken by our sequence (which is why the model is called "hidden"), but we have methods to calculate which path maximizes the overall probability of a given alignment of our target sequence to the model (i.e., to the family of proteins used to generate it).

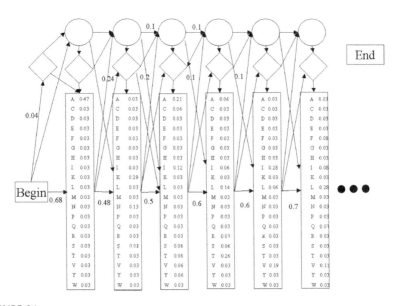

FIGURE 21

Graphical representation of the Hidden Markov Model derived from the multiple alignment shown in Figure 15.

Publicly available Hidden Markov Models have been built to represent several protein families, and they can be used to assess the probability that a newly determined sequence belongs to one of the families.

A major issue in computational biology is the identification of proteins homologous to a given input or query protein in a database or known protein sequence, and Hidden Markov Models are among the most powerful tools that can be used to make such identifications. The models are especially suitable for detecting distant relationships. Other methods are discussed below.

Database Searching

All known protein sequences and their functional annotations, when available, are stored in biological databases. A newly determined sequence can belong to a family of proteins of which one or more members have already been identified, characterized, and stored in the databases. Therefore, we must compare our protein sequence (the query) to each and every known protein sequence (the targets) to detect potential evolutionary relationships.

From a computational point of view, the problem is to align the query sequence to a very large collection of sequences and to sort them according to the score of their alignment with the input sequence. The method has to be fast and scalable because the number of available sequences is enormous and grows at an impressive rate (Figure 22).

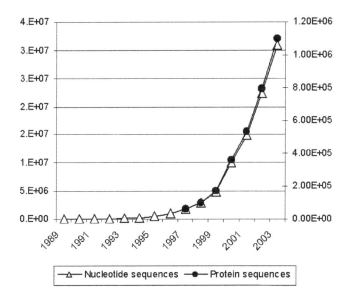

FIGURE 22

The growth of the nucleotide and protein sequence database. Note that the scales for the nucleotide and protein sequences (shown on the left and on the right, respectively) are different. (Data from http://www.ncbi.nlm.nih.gov and http://www.ebi.ac.uk.)

From a biological point of view, however, the problem is much more complex. What are we looking for? As we discuss in Problem 4, if our aim is to infer the structure of the query protein, we must obtain realistic alignments with homologous proteins, but if our aim is function assignment, we must distinguish between paralogous and orthologous sequences. We must also make sensible functional inference; that is, we must understand how much of what is known for an orthologous protein can be transferred to the query protein and, especially, the reliability of the functional annotations of the proteins in the databases.

The most commonly used methods for database searching are based on heuristics, and we must know their underlying assumptions because these assumptions have significant biological implications.

FASTA (Fast-All) is a sequence searching package that uses a multistep approach. It first finds sequences sharing exact short matches with the query protein by use of a lookup table. It then extracts sequences with a high number of short exact matches that can be part of the same alignment. Finally, it aligns the selected sequences with the query sequence by implementation of a dynamic programming algorithm.

Blast is another sequence searching package that searches for short exact matches between the query sequence and database sequences, which it tries to extend in both directions. It stops when the score is likely to be maximum.

A modified version of Blast, Psi-Blast, first runs a database search by Blast, collects and aligns the sequences likely to be homologous to the query, and uses those sequences to build a profile. The profile can be used to calculate the probability that a new sequence belongs to the multiple-sequence alignment as described before, and, thereafter, Psi-Blast uses the profile to search for other related sequences, collects new similar sequences, rebuilds a profile, and proceeds iteratively until no new sequences are found or until a user-defined number of iterations has been performed.

Any of the database search packages can provide a sorted list of matching sequences that putatively includes all sequences in the database that are homologous to the query. This list includes false positives (i.e., sequences that are not homologous to the query) and false negatives (i.e., homologous sequences that are identified as not homologous). These errors have different impacts on biological research.

The appearance of false negatives can be attributed to the assumption that homologous sequences necessarily include short exact matches and to the statistics used to assess whether a given score is significant from a biological point of view. The false negative problem is probably biologically less important because a new database search that uses shorter match lengths or, at worst, a brute-force approach that uses a full dynamic program algorithm on the whole database could, at least in principle, detect false negatives. In any case, the error does not propagate itself.

False positives are much more serious errors. They are related to the complexity of biological systems and cannot be solved by brute force. False positives not only can mislead biologists but also can, more importantly, cause error propagation in database functional assignment. Because of the speed at which new sequences are determined, the assignment of their function is mostly based on sequence similarity and, therefore, on the results of database searches, rather than on biochemical experiments. Thus, a false-positive error can be propagated, as newly determined sequences similar to the query sequence are assigned the same function on the basis of the detected similarity.

To assess the significance of a match, we should compare the observed score with the score expected by chance alone (i.e., with the background random distribution). An observed score significantly higher than what we expect by chance alone points to an evolutionary relationship. However, the nonrandomness of protein sequences makes the choice of the background distribution quite tricky. If we had a list of sequences certainly unrelated to the query, the results of a database search on this set of sequences would give us the background distribution, but such a list is very difficult to obtain. Therefore, we use the background distribution obtained on the same database by using as a query a sequence, or a set of sequences, unrelated to the query.

Blast calculates the background distribution by using as a query a sequence with the average composition of the database sequences. In other words, Blast searches the database many times with sequences that are generated randomly but have the average composition of the database and stores the obtained distribution of scores. When the user searches the database with a query sequence, each score obtained from the alignment with a database

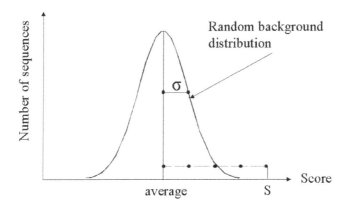

$$Z\text{-score} = (S\text{-average})/\sigma = 4$$

FIGURE 23
The Z-score is defined as the difference between the observed score and the average score of
the background distribution divided by its standard deviation. The higher the Z-score, the less
the likelihood that the match with SP-score is the result of chance alone. Here, we show the
Z-score for a Gaussian distribution.

protein sequence is compared with the random distribution of scores, and,
if it is statistically significantly higher (see Figure 23), the two sequences are
predicted to be homologous.

The advantage of this approach is that the expected random distribution
of scores can be calculated only once for a given database and a given scoring
system, but the significant drawback is that the composition of a query
sequence is not necessarily similar to the average sequence and, in some
cases, can be very different. Examples of the latter are sequences that contain
regions where a few amino acids are very frequent (i.e., the so-called low-
complexity regions). Especially if the very frequent amino acids are among
those that have a high score for exact matches, the risk is that any other
sequence that contains a large number of these amino acids will score very
high, even if unrelated to the query. Blast tries to solve the problem by
detecting and masking these regions (i.e., it does not use them to compute
the score). However, masking might increase the number of false negatives
because the presence of a low-complexity region in the query protein, which
is biologically relevant, is totally neglected in searching for homologs.

FASTA uses a different approach. It calculates the expected random distribution of scores by repeating the search on subsets of the original database
that are *generated from the query by reshuffling its sequence*. This approach is
costly in terms of computational time, but it decreases the risk of false
negatives for proteins containing low-complexity regions.

The other problem related to false negatives stems from the length of the
query and the database sequences. An exact match is more likely to be found
in a long sequence than in a short sequence. The background distribution
against which we compare our score depends upon the length of the

sequences. Very long protein sequences are usually indicative of the presence of more than one domain, and we would like to search the database with each domain independently because they can be derived from different ancestral sequences. However, detecting domain boundaries in a protein sequence is not easy. Several methods are available that are described in Problem 2, but we should mention here that their reliability is not yet satisfactory.

Reliability of Present Methods and Promising Avenues

When discussing the reliability of bioinformatics tools, one must take into account the biological question that the tools are trying to address. For sequence alignment and database search, the first distinction that must be made is whether these methods are used for the analysis of a single protein of interest or for high-throughput assignment of function or structure.

For a single protein, the reliability can be improved by a careful analysis of the results. A basic understanding of protein structure and function might be sufficient to assess whether the result can be trusted or whether the case at hand is borderline and more investigations are needed. For example, the distinction between orthology and paralogy can be based upon the observation of which amino acids are conserved and whether they include residues known or likely to be part of the active or functional site. A similarity limited to a domain can be immediately spotted and, for example, the alignment or database search can be repeated on fragments of the query protein.

High-throughput usage of the tools is much more complex. Ideally, one would prefer, for the functional assignment, to give a heavier weight to functionally important residues in the alignment. This preference is partially addressed by scoring schemes that take into account the conservation pattern of each residue in homologous families, but this approach is not always sufficient.

An important aspect to be considered is that a large body of information about specific proteins is available in the biochemical literature, and this information might include the results of mutagenesis or residue modification experiments on one of the proteins of the family. If these data could be taken into account in the alignment procedure, results would be much more biologically accurate.

We will return several times to the advantages of exploiting available experimental information on the proteins of interest and will discuss current techniques for doing so. Here, we mention only that a scoring system able to take into account information automatically extracted from the literature would be very helpful in this effort and with other bioinformatics problems as well.

Finally, if the three-dimensional structure of one of the members of the protein family is known, this valuable information should be, but only in a few methods is, used by the alignment algorithms. A position-dependent gap penalty as well as a joint probability of residue substitution for neighboring amino acids is expected to improve the results and also to help in the paralogous versus orthologous detection problem.

Suggested Reading

Lesk, A.M. *An Introduction to Bioinformatics*, 3rd ed., Oxford University Press, Oxford, 2005.

Durbin, R., Eddy, S., Krogh, A., and Mitchison, G. *Biological Sequence Analysis: Probabilistic Models of Proteins and Nucleic Acids*, Cambridge University Press, London, 1988.

Jorf, I., Yandell, M., and Bledell, J. *Blast: An Essential Guide to the Basic Alignment Search Tool*, O'Reilly and Associates, Sebastopol, CA, USA, 2003.

Orengo, C.A., Jones, D.T., and Thornton, J.M. Bioinformatics: Genes, proteins and computers, BIOS Scientific Publishers, Oxford, UK, 2002.

Eidhammer, I., Jonassen, I., and Taylor, W.R. *Protein Bioinformatics: An Algorithmic Approach to Sequence and Structure Analysis*, Wiley, New York, 2004.

Dayhoff, M.O., Schwartz, R.M., Orcutt, B.C. A model for evolutionary change, in *Atlas of Protein Sequence and Structure*, Dayhoff, M.O., Ed., Vol. 5, National Biomedical Research Foundation, Washington, D.C., 1978, pp. 345–358.

Smith, T. and Waterman, M. Identification of common molecular subsequences, *J. Mol. Biol.* 147, 195–197, 1981.

Pearson W.R. and Lipman D.J. Improved tools for biological sequence comparison, *Proc. Natl. Acad. Sci. USA* 85, 2444–2448, 1988.

Altschul, S. Gap costs for multiple sequence alignment, *J. Theor. Biol.* 138, 297–309, 1989.

Lipman, D.J., Altschul, S.F., and Kececioglu, J.D. A tool for multiple sequence alignment, *Proc. Natl. Acad. Sci. USA* 86, 4412–4415, 1989.

Altschul, S.F., Gish, W., Miller, W., Myers, E.W., and Lipman, D.J. Basic local alignment search tool, *J. Mol. Biol.* 215, 403–410, 1990.

Henikoff, S. and Henikoff, J.G. Amino acid substitution matrices from protein blocks, *Proc. Natl. Acad. Sci. USA* 89, 10915–10919, 1992.

Schneider T.D. and Stephens, R.M. Sequence logos: a new way to display consensus sequences, *Nucleic Acids Res.* 18, 6097–6100, 1990.

Gonnet G.H., Cohen M.A., and Benner S.A. Analysis of amino acid substitution during divergent evolution: the 400 by 400 dipeptide substitution matrix, *Biochem. Biophys. Res. Commun.* 199, 489–496, 1994.

Higgins, D.G., Thompson, J.D., and Gibson, T.J. Using CLUSTAL for multiple sequence alignments, *Methods Enzymol.* 266, 383–402, 1996.

Eddy, S.R. Hidden Markov models, *Curr. Opin. Struct. Biol.* 6, 361–365, 1996.

Park, J., Teichmann, S.A., Hubbard, T., and Chothia, C. Intermediate sequences increase the detection of homology between sequences, *J. Mol. Biol.* 273, 349–354, 1997.

Notredame, C., Higgins, D.G., and Heringa, J.T. Coffee: a novel method for fast and accurate multiple sequence alignment, *J. Mol. Biol.* 302, 205–217, 2000.

Problem 2

Predicting Protein Features from the Sequence

Introduction to the Problem

The pattern of secondary structure elements of a protein, the sites, if any, where it is posttranslationally modified, the cellular compartments where it resides, and many other functional features are specified by the amino acid sequence. The methods that we describe in this problem all have in common the idea of extracting rules from sets of proteins known to share a specific feature and applying them to the set of unknown cases. The task is to infer one or more rules from a training set composed of proteins sharing a given property. If the rules are sufficiently general, they can be used to predict the presence of the analyzed property in other proteins.

Function can be either deduced by the presence of a specific set of amino acids (a deterministic pattern) or by estimating the probability that the given sequence or subsequence belongs to the set of positive training examples (stochastic methods). In either case, we have a conservation problem if we only use a set of examples that share the property to be predicted (positive examples) and a classification problem if we also have a set of negative examples (i.e., proteins known not to share the property) available.

Deterministic Patterns

Protein function is often carried out by a limited set of specific conserved amino acids, and the remaining amino acids are responsible for allowing them to be properly positioned in three dimensions. This relation is, for example, the case for residues forming active sites, binding specific ligands, or being recognized by enzymes that catalyze posttranslational modifications. For these residues, the observed property might be attributed to the presence of specific amino acids in a certain relative position in the sequence;

that is, a diagnostic sequence pattern can be associated with a specific property. These patterns are called deterministic patterns.

We discussed an enzyme of major pharmacological interest in the Introduction, the protease from the hepatitis C virus. The identification of its function was possible because of the observation that its sequence contains a deterministic pattern, and this observation provided the impetus for years of study directed toward inhibiting the activity of the hepatitis C virus, which might, in the future, provide a cure for hepatitis C.

Proteases, as we mentioned, are enzymes that catalyze the cleavage of peptide bonds. The hepatitis C virus NS3 protein belongs to a class of proteases in which the amino acid that performs the catalysis is a serine; they are collectively known as serine proteases. A large evolutionarily related family of these enzymes is designated chymotrypsin-like, from the name of one of its members, and is very well studied. Several sequences, structures, and biochemical characterizations are available for members of this family. They are distinguished by the presence of three amino acids that are essential for catalysis: the serine that we already mentioned, a histidine, and an aspartic acid. The analysis of the amino acid sequences of chymotrypsin-like serine proteases shows that the relative order of these three amino acids in the sequence is conserved (histidine before aspartic acid before serine), although they are not necessarily in the same position with respect to the N- or the C-terminus (beginning and end) of the protein. Catalysis is achieved by the enzyme through the stabilization of a high-energy reaction intermediate that contains a negatively charged oxygen atom. During the reaction, this atom is positioned in a pocket of the protein formed by two glycine residues or by a glycine and a serine immediately adjacent to the catalytic serine. Furthermore, the catalytic serine is always preceded by a glutamic or aspartic acid. These observations can be summarized by saying that chymotrypsin-like serine proteases contain the pattern [DE]-S-G-[GS].

The presence of this pattern in the sequence of the hepatitis C virus led to the identification of the function shortly after the sequence was available and prompted many efforts to inhibit its activity to obtain a drug against the disease.

In general, we can say that a pattern such as the one shown above can be defined as

$$P = p1,...pN; \; pi \in \Sigma \tag{1}$$

where Σ is the alphabet containing the 20 amino acid symbols. We can allow the inclusion of regions with no definite constraints and of variable length in the pattern. For example, the pattern D-X(1,4)-[L,I]-X-[D,E] describes a subsequence formed by the amino acid aspartic acid (D), followed by between 1 and 4 amino acids, followed by either a leucine (L) or an isoleucine (I), followed by any amino acid, followed by either an aspartic acid (D) or a glutamic acid (E).

This pattern is equivalent to rewriting Equation (1) as

$$P = p_1...p_n, \ p_i \in \Sigma \cup X, \text{ where } X = \{x(n1,n2) \mid n1 < n2 \in N\} \qquad (2)$$

and x is any amino acid symbol.

We can also include a notation to express the exclusion of 1 or more amino acids; for example, {L,I} can indicate any amino acid, except a leucine (L) or an isoleucine (I). This notation is used by a database of patterns called PROSITE, which collects known deterministic sequence patterns for a variety of functions.

Deterministic sequence patterns can be learned by analysis of the sequence-structure-function relationship in a class of proteins, as in our chymotrypsin-like serine protease example, or the patterns can be derived by comparison of the sequences of a set of proteins that share the target property.

If we have a set of sequences of proteins that are known to share a common property, we can enumerate a set of possible patterns, calculate how well each pattern fits the examples on the basis of a predefined fitness function, and select the patterns with highest fitness. The most natural way to enumerate the patterns is to determine the length of the pattern and use as the initial set all patterns of the given length present in the set of examples; that is, all substrings of the predefined length present in the protein sequences that share the property to be predicted. At the end of the procedure, more than one pattern can be combined to obtain the optimal pattern, and this method guarantees that, up to some limited size, the best patterns can be found almost regardless of the total length of the examples.

Another possibility is to construct a multiple-sequence alignment of the sequences, identify conserved amino acids, and cluster them together to derive a pattern. For example, the alignment in Figure 14 can be used to derive the pattern G-G-L-G-X-L-A-X(4,5)-S.

The observation that the optimal multiple sequence alignment cannot be exactly computed, as we discussed, implies that these latter methods, although able in principle to find patterns of any length, must be based on heuristics and cannot be guaranteed to find the optimal solution. A natural extension of the conservation problem (where we assumed to have only positive examples) is to devise a score function that takes into account not only the presence of the pattern in the set of positive examples but also its absence in the set of negative examples.

Stochastic Patterns

A deterministic pattern lists the amino acids required in certain positions of a protein sequence. A stochastic pattern reports the probability that one amino acid occupies a certain position. We have already seen an example of a stochastic pattern in Problem 1, the profile that can be derived from a set of aligned sequences. As we discussed, a profile is a matrix in which each column represents a position in the sequence, and each row represents one of the 20 amino acids. Each cell contains the probability that the position that corresponds to the column is occupied by the amino acid that corresponds to the row. The probability values are estimated from the observed frequencies in the set of known examples. Given a protein sequence, this contingency table can be used to calculate the probability that the sequence belongs to the set of positive examples. The obtained value is then compared with that expected by chance alone, which can be derived from the set of negative examples or from an estimated background probability distribution. Let us reiterate here that protein sequences are not random sequences of 20 equally probable amino acids. Therefore, computing the background distribution is not trivial and is the most common pitfall in prediction methods.

A profile is a rather simple statistic model that describes a set of aligned sequences. As we mentioned previously, a more sophisticated probabilistic method is the Hidden Markov Model. We can construct a Hidden Markov Model from the alignment of functionally important regions of related sequences and, given a new sequence, use the model to evaluate the probability that the new sequence fits the pattern. This feature is the basis of the PFAM database that collects Hidden Markov Models for several functional domains and that can be used to obtain an estimate of the probability that a newly determined protein sequence belongs to one of the database families.

Other automatic learning methods, such as neural networks or support vector machines, are equally popular for feature prediction. We briefly describe the former because it forms the basis of several commonly used methods.

A neural network is an assembly of neurons, which we represent as nodes, with one or more incoming connections that we call input and an outgoing connection that we call output (Figure 24).

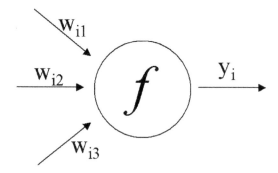

FIGURE 24

A neuron with three incoming connections (representing the input values) and one outgoing connection, which is some function of the input values, from which we can collect our output.

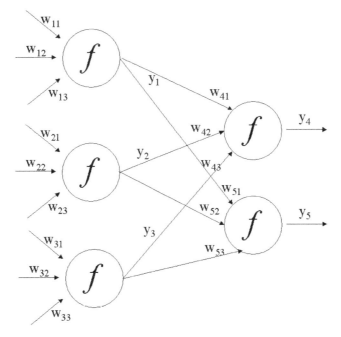

FIGURE 25
Connecting neurons. The output of three neurons becomes the input of other neurons, from which we can collect the output that is a function of the input values and of all the weights assigned to the input connections.

The output is connected to the input through some, usually simple, arithmetic operations, and each node contributes to the output according to a variable "weight." In physiologic terms, we say that the neuron "fires" if the output is above a given threshold and does not fire if the output is below the threshold. We can connect several neurons, in an almost unlimited number of ways, so that the output of some neurons represents the input of others (Figure 25).

The power of neural networks is that, during what we call the learning phase, we can use examples for which we know the answer (i.e., whether they share the feature to be predicted or not) as input, and the algorithm modifies the values of the weights to maximize the probability that the output matches the known answer. In the example shown in Figure 25, we can optimize the weights so that the value in y_4 is higher than that in y_5 when the positive examples are used as input and lower than that in y_5 when the negative examples are used. The neural network with the weights derived during the learning phase can subsequently be used to calculate the answer for an unknown example. In other words, the network "maps" the input values to one or more output values according to the rules that it has learned from known examples.

Most neural network–based methods, even if their aim is to provide a prediction on a single target amino acid, use as input a segment of the protein sequence that includes the target amino acid and a predefined number N of amino acids before and after the target, to take into account context effects, and use an input node for each of the amino acids of the input region. The segment is moved along the protein sequence so that each amino acid, with the exception of the first N and the last N, is used as the target.

The input subsequences of amino acids must be numerically encoded. If we use a single sequence as input, we usually assign a different binary code to each amino acid. We could use 5 bits to encode all 20 amino acids ($2^5 = 32$), but this method would introduce a problem. The amino acid encoded by 01111 would be at a greater "distance" from the amino acid encoded by 00001 than from the amino acid encoded by 01010. Therefore, we would be giving misleading information to our automatic learning machinery. This problem can be easily avoided by the use of 20 bits and a different bit set to 1 for each amino acid, while the other bits are set to 0 (sparse coding).

The output of the neural network is a set of numerical values that refer to the central amino acid and are associated to the property that we want to predict. For example, we can have three output numbers (or nodes as we call them), each associated to a type of secondary structure (α, β, or other), and the amino acid will be assigned to the secondary structure for which the output node has the highest value or a value higher than a predefined threshold.

If we know that the property we want to predict is conserved during evolution—and secondary structure is one such property—we can take advantage of this knowledge and use as input a profile derived from the multiple-sequence alignment of members of the family. This method is indeed the most commonly used method of encoding the input for protein bioinformatics neural networks (Figure 26).

The set of examples used in the training phase is crucially important. They should contain sufficient information to allow the parameters to be optimized, but care should be taken in their selection to avoid the situation in which the parameters merely reflect the input sequences and are, therefore, unable to generalize. For example, if we want to predict secondary structure, we should be very careful that no protein in the training set is clearly homologous to any protein of the testing set because we know that homologous proteins have the same overall structure. If the network learns how to "recognize" proteins that belong to the same family, the prediction for proteins of the testing set homologous to proteins of the training set will be more accurate than predictions for unrelated proteins, and we will overestimate the accuracy of our network.

Several widely used methods are based on neural networks. They are used for prediction of specific signals for posttranslational modification, cellular localization, secondary structure, accessibility to solvent, and other features.

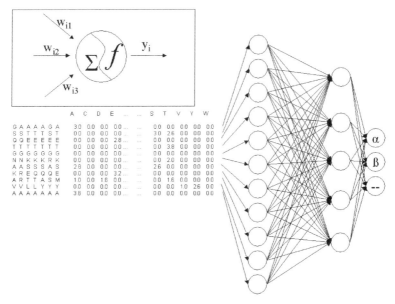

FIGURE 26
A neural network for the prediction of secondary structure. The input is a profile derived from a multiple-sequence alignment of homologous proteins shown here with each sequence running to bottom, and the output is one of three states: α, β, or neither. The output refers to the central amino acid of the input window.

Specificity and Sensitivity of a Feature Prediction

How good is a feature prediction method? Intuitively, we would like our method to detect all cases that share the analyzed property and none of the others. In other words, we would like the method to have as few false negatives (i.e., cases in which the method fails to detect an existing feature) and false positives (i.e., cases in which the method predicts a nonexistent feature) as possible. Clearly, the two values are correlated: reducing the stringency of the method reduces the number of false positives but is likely to increase the number of false negatives.

The diagnostic power of a method can be expressed in terms of its specificity and sensitivity. These two parameters are defined as

$$\text{sensitivity} = F^+/(F^+ + f_n)$$

$$\text{specificity} = F^-/(F^- + f_p)$$

where F^+, F^-, f_n, and f_p are the number of correctly identified positive cases, the number of correctly identified negative cases, the number of false negatives, and the number of false positives, respectively.

The correlation coefficient is one of the measures of the diagnostic power of the method that combines both concepts:

$$C = \frac{F^+ \bullet F^- - f_p \bullet f_n}{\sqrt{(F^+ + f_p) \bullet (f_p + F^-) \bullet (F^- + f_n) \bullet (f_n + F^+)}} \qquad (3)$$

If the number of false positives and false negatives is zero (that is, if the method is infinitely sensitive and specific) we have

$$C = \frac{F^+ \bullet F^-}{\sqrt{(F^+) \bullet (F^-) \bullet (F^-) \bullet (F^+)}} = 1 \qquad (4)$$

The value of C decreases towards 0 as the number of f_p and f_n increase.

The ROC Curve

Another very commonly used description of the reliability of a method is the so-called ROC (receiver-operating characteristic) curve. It was developed during World War II as a way to measure the ability of radar operators to distinguish between noise and real radar signals. The problem this approach is designed to solve is that sensitivity and specificity are correlated. Suppose we have two methods of predicting whether or not a site is a serine phosphorylation site, and the predictions are based on some stringency value that is differently defined in the two cases. How can we compare the accuracy of the methods?

Presumably, increasing the stringency in each method reduces the number of false positives and increases the number of false negatives. The most natural solution is to evaluate the methods on a known data set and see how many times they correctly predict a phosphorylation site, while predicting the same fraction of false negatives, or vice versa. The best way to visualize this approach is to plot the true-positive fraction as a function of the false-positive fraction.

Another important advantage of the ROC curve is that the area under the curve is related to the accuracy of the method (see Figure 27). Two methods are commonly used for computing the area: a nonparametric method based on constructing trapezoids under the curve and a parametric method that utilizes a maximum-likelihood estimator to fit a curve to the data points.

The Prediction of Protein Domain Boundaries

The definition of a domain is rather fuzzy. When we study a protein structure, the detection of the presence of several domains is quite intuitive. As

ROC curves

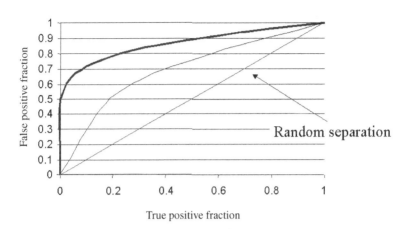

FIGURE 27

ROC curves for three hypothetical prediction methods. The accuracy of a method depends on how well the test separates positives and negatives and is measured by the area under the ROC curve. An area of 1 represents a perfect test; an area of 0.5 represents the results for a random separation.

we discussed in Problem 1, they represent globular, compact regions of the protein structure with relatively more contacts within themselves than with the rest of the structure, but the precise boundaries are often difficult to calculate even when the structure is available because methods differ and, often, manual inspection is required (Figure 28) (see color insert after page 40). This observation means that we do not have a clean and reliable set of positive examples on which we can train our methods, which makes the identification of domain boundaries from the sequence of a protein even more complex. Nevertheless, methods are continually being developed because they are of paramount importance for sequence analysis and for three-dimensional structure modeling, as we will see in Problem 3.

By and large, methods can be subdivided into two groups: those that only use the amino acid sequence of the target protein and those that take advantage of the similarity of its sequence with other proteins on the database.

The domain guess by size (DGS) method is based on a probability distribution derived from an analysis of the length and distribution of domains in known protein structures. Its reliability is not very high, but it is a useful guide for human expert assignment. Another method relies on the assumption that the larger contribution to entropy loss that occurs during protein folding is the result of the restriction of the degrees of freedom of protein side chains, and that this loss must be compensated by interresidue favorable interactions. Amino acids with a high number of possible conformations (and, therefore, "in need" of more interactions) are more likely to be found within domains than between them.

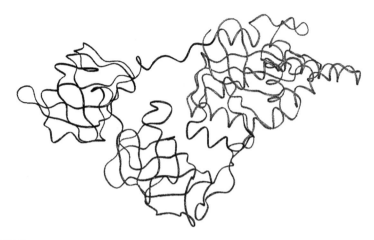

FIGURE 28

The structure of the elongation factor–1 from *Sulfolobus solfataricus*, a protein involved in mRNA translation. The three domains are connected by long stretches of amino acids, and some regions are not clearly packing against one or the other domain. Consequently, precisely defining the domain boundaries within the regions colored in black is difficult.

The SnapDRAGON tool produces several hundred putative, three-dimensional models of the target protein and detects its domains by averaging the results of the predictions. This method is very computationally intensive and not very useful for large-scale analysis. A less computationally intensive method, DomSSEA, predicts the secondary structure of the target protein and maps the predicted sequence of helices and sheets with those observed in known protein domains. It thereby evaluates the probability that a given set of secondary structure elements is sufficient to fold into a domain.

Unfortunately, none of these methods is sufficiently reliable to predict the domain structure of an unknown protein in a completely automatic fashion, and this shortcoming is a serious obstacle to solving many computational biology problems.

Methods based on database searches are generally less applicable, but more reliable. The detection of matches that only span a region of the protein under examination is obviously a strong indication of the presence of a domain, as is the detection of a match to a known domain, for example, one of those stored in domain databases such as PFAM.

Domains in proteins are not necessarily contiguous in sequence. The amino acid chain can start folding into a domain, make an excursion and form another domain, and then come back to complete the first domain (Figure 29) (see color insert opposite this page). These cases are difficult to handle, and, indeed, no satisfactory method is available to detect these cases on the basis of the protein amino acid sequence alone.

The field of bioinformatics is evolving, and, although several attempts to combine existing prediction methods to achieve more satisfactory results

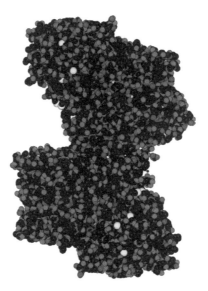

COLOR FIGURE 1
An all-atom representation of a protein structure determined by X-ray crystallography. This protein is an enzyme, glycogen phosphorylase from rabbit muscle, and its code in the Protein Data Bank is 1ABB. Atoms are colored according to a commonly used scheme: carbon is black, nitrogen is blue, oxygen is red, and sulfur is yellow.

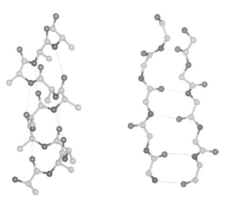

COLOR FIGURE 6
The backbone atoms of an α-helix and of two β-strands are depicted above. The strands, pairing via hydrogen bonds (dotted lines) form a β-sheet.

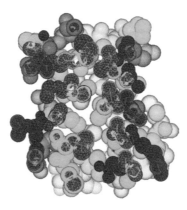

COLOR FIGURE 8
A section of the structure of SH3, a small module found in many proteins, where it acts as an adapter to recruit other proteins. The green hydrophobic amino acids are more frequent in the inside than on the outside of the molecule.

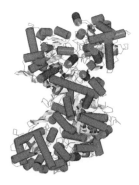

COLOR FIGURE 9

The structure of glycogen phosphorylase once again. This time helices and strands are shown as cylinders and arrows.

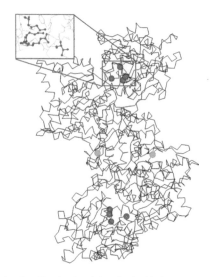

COLOR FIGURE 10

The active site of glycogen phosphorylase. The phosphorylation of serine 14, shown as a green ball, triggers a conformational change in the protein.

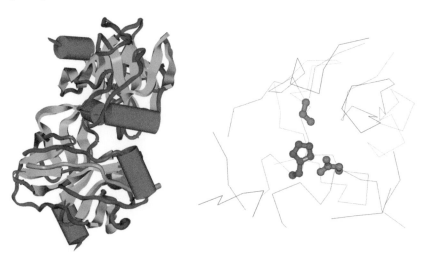

COLOR FIGURE 11

The structure of the protease of the hepatitis C virus (PDB code: 1NS3).

```
Sequence 1       AL KTLNYDF DHLVEMESDAGLGNGGLGRLAACYLDSMATLAV
Sequence 2       VMKEFDLDLNEI I EQEPDPGLGNGGLGRLAACFLDSL ASLEV
Common residues   K      D       E  E  D  GLGNGGLGRLAAC  LDS  A  L  V


Sequence 1       AL KTL NYDF DHLVEMESDA GL GNGGLGRLAACYL DSMAT LAV
Sequence 4       AYF SAEF GVHETL PI YS- GGL- - - - - GVLAGDHVKSA SDLNL
Common residues A                  S     GL         G  LA        S


Sequence 1       AL KTLNYDF DHLVEMESDAGLGNGGLGRLAACYL DSMAT LAV
Sequence 2       VMK EFDLDLNEI I EQEPDPGLGNGGLGRLAACF L DSL ASL EV
Sequence 3       AL MDLGFKLEDLYDE ERDAGLGNGGLGRLAAC- MDSL ATCNF
Sequence 4       AYF SAEFGVHETL PI YS- - - - - - GGLGVLAGDHVKSA SDL NL
Common residues                          GGLG  LA          S
```

COLOR FIGURE 14

The first panel shows a pairwise alignment between two evolutionarily related sequences. The two sequences are very similar and, therefore, easy to align. However, their similarity is such that most of their residues are identical, and, therefore, a determination of which are really important for function is difficult. The second panel shows the alignment of two distantly related sequences. This alignment is more useful in highlighting important residues but is more ambiguous. The third panel shows a multiple alignment between four sequences. This alignment is a better compromise because it is more reliable, and the conserved residues are easier to detect. The last part of the figure shows a useful graphical representation of a multiple-sequence alignment as a stack of symbols, one stack for each position in the sequence. The height of symbols within the stack indicates the relative frequency of each amino acid at that position.

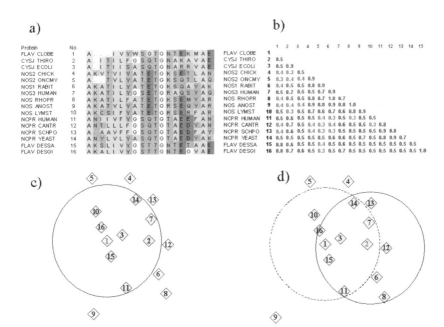

COLOR FIGURE 15

"Hopping" in sequence space. (a) A partial alignment of a protein sequence with a set of evolutionarily related proteins. (b) A matrix in which the percentage of different amino acids between each pair of sequences is charted. If we make the simplifying assumption that a difference lower than 0.4 is statistically significant, we can then conclude that sequences 2, 3, 7, 10, 11, 14, 15, and 16 are likely to be evolutionarily related to sequence 1. On the other hand, because homology is transitive, proteins evolutionarily related to sequence 2 are also related to sequence 1. We can highlight all sequences that are statistically more similar than expected to sequence 2 and, thereby, add sequences 6, 8, 12, and 13 to the family.

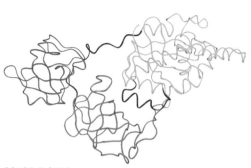

COLOR FIGURE 28
The structure of the elongation factor–1 from *Sulfolobus solfataricus*, a protein involved in mRNA translation. The three domains are connected by long stretches of amino acids, and some regions are not clearly packing against one or the other domain. Consequently, precisely defining the domain boundaries within the regions colored in black is difficult.

COLOR FIGURE 29
A discontinuous domain in the RNA 3'-terminal phosphate cyclase enzyme from yeast.

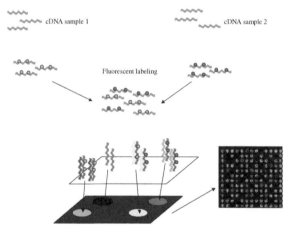

COLOR FIGURE 33
Scheme of transcriptomics experiments. cDNA molecules derived from different samples are treated with two different dyes, mixed, and made to hybridize with a previously prepared microarray. If the cDNA complementary to that contained in one of the microarray spots is present in both samples, the spot will be yellow. Red or green indicates that the molecule is only present in one of the two samples, and black indicates that it is absent in both.

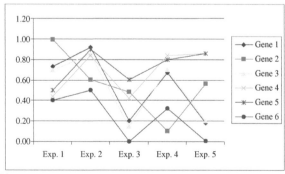

COLOR FIGURE 36
Graphical representation of the level of expression of the first six genes shown in Figure 34.

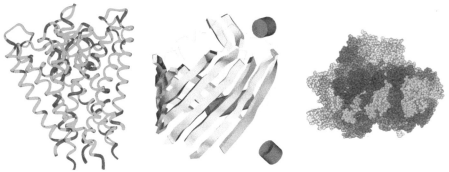

COLOR FIGURE 51
Examples of integral membrane proteins: a channel protein, a porin, and the photoreaction center.

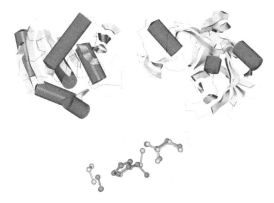

COLOR FIGURE 53
Comparison of the active sites of a protease and a lipase. The topology of the two proteins is very different (top), but the residues in their active site can be almost perfectly superimposed.

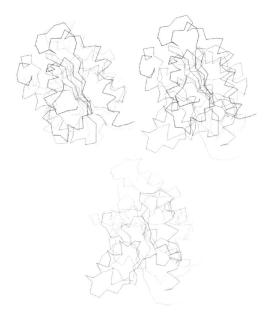

COLOR FIGURE 55
Different superpositions of evolutionarily related proteins: both superpositions have an rmsd of about 3 Å, but the one on the right includes 64 residues and the other includes only 36.

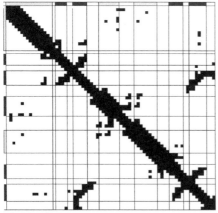

COLOR FIGURE 56
The distance matrix for an SH3 domain (extracted from the PDB entry 1RUN), the structure of which is shown on the right. Cells containing distances lower than 8 Å are filled. The secondary structure of the protein is shown as gray (helices) and black (strands) bars in the first column and in the first row.

COLOR FIGURE 58
Hemoglobin is the oxygen transporter. The protein is α-helical and is formed by four chains, two α-chains and two β-chains bound by noncovalent interactions. It contains one heme group per chain.

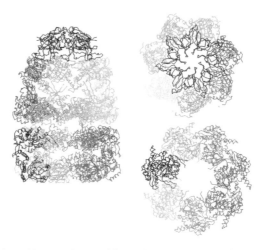

COLOR FIGURE 60
GroEL and GroES. The chains of GroEL are shown in different colors. GroES is the "cap" shown in black. The two images on the left are the same protein seen from top and bottom, respectively.

COLOR FIGURE 61

The structure of a portion of the rhinovirus coat bound to an inhibitor (left) and to an antibody (right). The virus has an icosahedral shape. Each of the triangular faces is formed by three protein chains shown in different colors.

COLOR FIGURE 62

The structure of a fragment of an antibody (PDB id: 3HFL) bound to an antigen. The contact region of the antibody is shown in green dots, and that of the antigen (lysozyme) is shown by a solid blue surface.

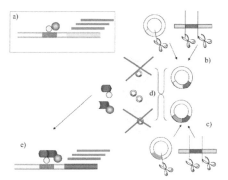

COLOR FIGURE 65

A simplified view of the rationale of a yeast two-hybrid experiment. The gal4 system is shown in (a). The gal4 protein, composed of a DNA-binding domain and an activation domain, activates the transcription of the downstream gene. Two genes coding for two interacting proteins are cloned and fused to each of these domains as shown in (b) and (c). The plasmids are transfected into a yeast cell population. Only in cells that contain both fusion proteins, the spatial proximity between the DNA-binding domain and activation domain is reconstituted, which leads to transcription of the downstream gene. If the original gene is replaced by a gene essential for cell survival, only cells that contain both fusion proteins can survive. This system can, therefore, be used to determine whether two proteins interact. Moreover, if the protein of interest is fused to one of the domains and a population of proteins is fused to the other domain, the method allows the identification of which proteins in the population, if any, interact with the protein of interest.

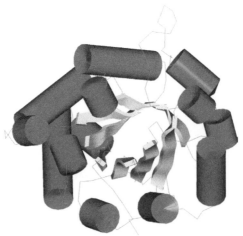

COLOR FIGURE 73
A TIM barrel (PDB id: 8TIM).

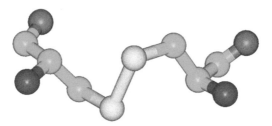

COLOR FIGURE 75
A disulfide bridge.

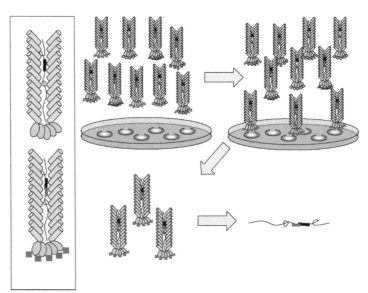

COLOR FIGURE 80
Phage display experiments. An oligonucleotide cloned upstream of the gene for the pV protein of the bacteriophage (shown as a gray oval) is exposed on the surface of the phage. If the cloned sequence is random, each phage will display a different peptide sequence. The phage population can be made to interact with a target protein. Only phage that contain a peptide that binds to the target protein will be retained, whereas the others can be washed away. The DNA sequence of the selected phage will directly reveal the sequence of the interacting peptide. The pVIII protein, shown as a gray cylinder, is also suitable for displaying foreign peptides on the surface.

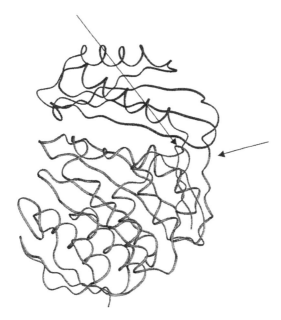

FIGURE 29
A discontinuous domain in the RNA 3'-terminal phosphate cyclase enzyme from yeast.

have been made, new ideas are needed in this area. This need is also indicated by the establishment of community-wide initiatives for blind evaluation of the reliability of domain detection methods, in which domain boundary predictions for proteins of unknown structure are collected and stored. Subsequently, when the protein structure is experimentally determined, the sequence-based predictions are compared with the structure-based assignment of the domain boundaries.

Reliability of Present Methods and Promising Avenues

At least in principle, the more sequences we collect and store in our databases, the more likely that we can detect regularities associated with their sequence features. Unfortunately, many of the entries in our collections come with no or only partial functional assignment, and, therefore, instead of contributing to a better definition of features, they only increase the gap between the known examples and those for which our knowledge must be extrapolated. This issue is not trivial. Accurate manual functional assignments such as those provided by curated databases lag behind because they are time- and labor-consuming and cannot keep pace with the speed at which new sequences are determined. As we mentioned, functional assignment of newly determined protein sequences is often made on the basis of their similarity with known cases, most of which have been assigned via the same route. In practice, the proteins that have been experimentally analyzed are but a very tiny fraction of our collection, and we must, therefore, be aware of the methods used to assign the functional annotations and of their reliability.

Databases such as PROSITE report the number of false negatives and false positives detected by each of the patterns, and the results of automatic learning methods on a properly selected test set can be used to assess their reliability. However, a problem we should not underevaluate is how representative of the biological diversity is the set of our known examples.

The proteins that have been experimentally annotated were not chosen at random. They are proteins of specific pharmacological interest, proteins that are easy to handle in the laboratory, or proteins from organisms that can be grown in culture or that represent suitable model systems. These proteins are most likely not representative of the menu of possibilities. We may discover that our methods have been optimized for the limited, nonrandomly selected set of examples available today and will in the future need to be reoptimized.

Programs are underway for sequencing nucleic acids found in different environments, such as the Sargasso Sea, and they will be instrumental in determining how much of our knowledge can be extrapolated to a completely different set of data, which is likely to be disjoint from what we have analyzed so far. A reassessment must be done of our knowledge. Have we really already catalogued about 60% of the known protein families? Is our estimate of the different types of protein architectures present in nature correct?

Suggested Reading

Hofmann, K., Bucher, P., Falquet, L., and Bairoch A. The PROSITE database, its status in 1999, *Nucleic Acids Res.* 27, 215–219, 1999.

Henikoff, J.G., Greene, E.A., Pietrokovski, S., and Henikoff, S. Increased coverage of protein families with the blocks database servers, *Nucleic Acids Res.* 28, 228–230, 2000.

Bateman, A., Birney, E., Durbin, R., Eddy, S.R., Howe, K.L., and Sonnhammer, E.L. The Pfam protein families database, *Nucleic Acids Res.* 28, 263–266, 2000.

Wheelan, S.J., Marchler-Bauer, A., and Bryant, S.H. Domain size distributions can predict domain boundaries, *Bioinformatics* 16, 613–618, 2000.

George, R.A. and Heringa, J. SnapDRAGON: a method to delineate protein structural domains from sequence data, *J. Mol. Biol.* 316, 839–851, 2002.

Marsden, R.L., McGuffin, L.J., and Jones, D.T. Rapid protein domain assignment from amino acid sequence using predicted secondary structure, *Protein Sci.* 11, 2814–2824, 2002.

Venter, J.C., Remington, K., Heidelberg, J.F., Halpern, A.L., Rusch, D., Eisen, J.A., Wu, D., Paulsen, I., Nelson, K.E., Nelson, W., Fouts, D.E., Levy, S., Knap, A.H., Lomas, M.W., Nealson, K., White, O., Peterson, J., Hoffman, J., Parsons, R., Baden-Tillson, H., Pfannkoch, C., Rogers, Y.H., and Smith, H.O. Environmental genome shotgun sequencing of the Sargasso Sea, *Science* 304, 66–74, 2004.

Problem 3

Function Prediction

Introduction to the Problem

The objective of protein bioinformatics is function prediction. Thus, all we have said so far and all we will discuss in the next problems is about function prediction. Yet, some specific aspects must be discussed separately, and they are the subject of this problem.

We described several methods of obtaining information about protein function. The detection of evolutionary relationships is one method, as is finding characteristic sequence patterns. As we will see in subsequent problems, the experimental determination or the prediction of a protein structure can also be very powerful. However, experimental techniques, made possible by the knowledge of the genomic sequences, produce a vast amount of data, the analysis of which requires the development of bioinformatics tools. Before we describe the state of the art of these tools, we must address a very complex problem: how do we define and catalog the complex concept of biological function?

The Definition of Biological Function

Biological function has several "dimensions." For example, the functional attributes of the protease of the hepatitis C virus can be any of the following:

- An enzyme catalyzing a chemical reaction
- A protein involved in HCV infection
- A protein expressed in liver cells

All three descriptions are accurate. However, they reflect different categories of functional attributes: molecular, biological, and cellular, respectively.

Furthermore, molecular function can be defined at several levels of detail. The protein is an enzyme, in particular, a hydrolase (it breaks a chemical bond), a peptidase (the specific bond cleaved is a peptide bond), an endopeptidase (because it cleaves an internal bond in a peptide chain), a serine-type endopeptidase, or a chymotrypsin-like serine-type endopeptidase. We can go even more in depth by saying that the protein performs its function by utilizing the OH group of a serine, with the amino group of a histidine acting as a general base and a general acid, which, in turn, is oriented by the presence of a charged aspartic acid. We could continue by describing how the reaction intermediate is stabilized by the protein and how the substrate is recognized. How we define function is important because when our protein sequence is deposited in the database, the experimentalist or the database expert will add functional annotation. Which functional attributes will he or she assign, and at what level of detail?

Another aspect is lack of standardization, which can create serious problems for automatic tools. Biologists are not fond of standard nomenclature, and most of the terms they use have a historical justification. This practice significantly affects the assignment of specific names to specific proteins, as we will discuss later.

The situation is somewhat clearer for enzymes, where a standard nomenclature, the so-called enzyme classification or EC scheme, has been devised as described in Figure 30. Some databases, such as SwissProt, use a controlled vocabulary to provide functional annotations. A scientific consortium, the Gene Ontology, or GO, Consortium, has begun a commendable effort to define a standard vocabulary for function. This effort has already had beneficial effects.

The Function Vocabulary

Protein sequence databases contain both the amino acid sequence of the protein and some annotations, including definitions of its function. In some cases, the annotations are free-text entries. The structure of the SwissProt database is more controlled, and, therefore, it has a higher added value. The annotation data of SwissProt include the function (based on a controlled keyword vocabulary), the location and type of posttranslational modifications, the location of domains and specific binding sites, the secondary and quaternary structure of the protein, a list of proteins similar to the entry, a list of diseases associated with malfunction of the protein, and links to related entries in other databases. Perhaps more importantly, SwissProt annotations are periodically reviewed and updated by experts on specific groups of proteins. Manual verification is clearly the ideal way to ensure correctness

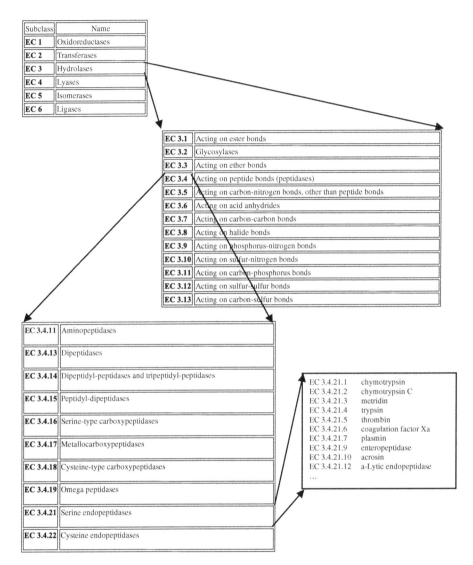

FIGURE 30

The EC nomenclature. It is based on the recommendations of the Nomenclature Committee of the International Union of Biochemistry and Molecular Biology, and is a hierarchical classification based on a four-digit scheme.

of the annotations, but we cannot ignore the fact that the approach is not scalable. The number of proteins for which annotations are available lags behind the number of proteins of known sequence by orders of magnitude, and the situation will become worse as time passes.

The GO tool is not a database of gene sequences or a catalog of gene products. Its goal is to give consistent descriptions of gene products in different databases. The descriptions are based on three structured,

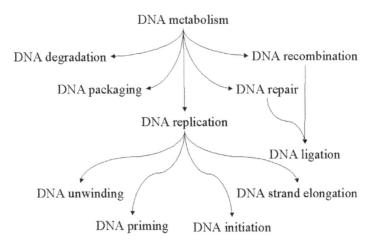

FIGURE 31
An example of the Gene Ontology structure.

controlled vocabularies (ontologies) that describe biological processes, cellular components, and molecular functions in a species-independent manner. GO terms are now used by other databases, and this acceptance facilitates their uniform usage.

Molecular function describes activities, such as catalytic or binding activities, at the molecular level, performed by individual gene products or by complexes of proteins. A biological process is a process with more than one distinct step and is the result of the action of one or more ordered assemblies of molecular functions. A cellular component is a component of a cell that is part of some larger entity, which may be an anatomical structure or a gene product group. A gene product can have more than one molecular function, can be used in more than one biological process, and can be associated with one or more cellular components. Therefore, GO terms represent activities rather than the molecules that perform the functions, and the structure cannot be hierarchical because a "child" (more specialized term) can have many "parents" (less specialized terms). In this case, the topology of the organization is called a directed acyclic graph (Figure 31).

Protein Names

Giving a protein a name that has something to do with its function and with the organism it belongs to would be convenient, but as we mentioned, biologists are not fond of standards. Sometimes proteins are named after their molecular weight (p53, a very important protein involved in DNA repair and whose malfunction is among the most common causes of cancer, weights about 53 kDa), after the scientists who discovered them (Huntington disease is named after its discoverer, and the protein later found to be involved in the disease has been named huntingtin). In some cases, the name of a protein reveals the mood of its discoverer (the YAKK protein is the acronym for Yet Another Kinase Kinase). Perhaps the most uninformative protein names can be found among the *Drosophila* (the fruit fly) ones. A blind mutant has an impaired cell function listed as the seventh function. The gene was, therefore, named sevenless, which is in itself not very informative, and other genes have been named after it: *dos* (daughter of sevenless), *sos* (son of sevenless), and *bos* (bride of sevenless). The *eag* gene owes its name to the abnormal behavior of anesthetized flies; it comes from "Ether A Go-Go." The same behavior can derive from mutations in other genes: Shaker (*Sh*) and Hyperkinetic (*Hk*). A gene responsible for difficulty in generating off-spring has been named Tudor. A gene that causes defects in structures called poles in the fly has been named "Scott of the Antarctic." Sometimes a new additional function is discovered for a known protein, and the protein is endowed with more than one name, often equally uninformative.

Some of the stories behind these names are interesting and reveal fascinating aspects of the scientific process that led to the discovery of the proteins' function, but these names are not useful for designing automatic tools to correlate the function of different proteins. They also have disastrous effects on attempts to retrieve information from the scientific literature.

Text Mining

Whether the route to function discovery is computational or experimental, the end result is, by and large, a scientific publication. These publications are stored and indexed in systems such as PubMed, and at least the abstracts of the scientific contributions are available in computer-readable form. All that is needed is some automatic system for extracting the functional information from the enormous amount of scientific literature. Not surprisingly, efforts to this end are flourishing. Many systems are already available, some of which focus on interactions or on the reconstruction of pathways, whereas others try to extract functional information on proteins or, in general, gene products.

The first problem to be faced in developing these systems is the identification of protein and gene names. These names can be long, compound names, and often, a different name for the same protein is used even within the same article. These names can also be common English words, as we discussed earlier. One solution is to use a dictionary extracted from one or more of the databases. Another solution is to look for properties such as the occurrence of uppercase letters, numbers, or special endings. These approaches can reach specificity and sensitivity of 70% to 80% (i.e., systems are able to recover 70% to 80% of the protein names present in the set of data, and 70% to 80% of the recovered names are correct). Hopefully, the increasing popularity of these methods and the exceptional advantages they confer will prompt experimentalists to be more careful in selecting naming schemes.

Once the names have been identified, the task is to determine the meaning of the sentence in which the names appear. This determination can be made by use of domain-specific grammar; that is, words and phrases that are commonly used in specific fields. In protein science, for example, such words and phrases could be "substrate is," "activity is," "interacts with," "is phosphorylated by," "is involved in," and "binds to."

Alternatively, methods can be used that were developed in the computational linguistics (or natural language processing) field. Workers in this field are making substantial progress in designing systems to perform such tasks as text analysis. For example, extracted phrases from an article or book, when shown to a human reader, seem to summarize the content.

The following is an abstract of a paper that describes the assessment of the results of a large-scale evaluation of methods for protein structure prediction:

> *ABSTRACT* — This report describes the assessment of the homology-based predictions submitted to the 5th edition of the Critical Assessment of Methods for Protein Structure Prediction (CASP5) experiment. We assessed the ability of the methods to predict the overall fold, the portions of the structure that differ substantially between the target protein and its closest structural homologue and the conformation of the side-chains. We also compared the results with those obtained in previous editions of the experiment and derived some general conclusions about the state of the art of comparative modeling methods and their usefulness for experimentalists.

The average frequency of each word in the abstract is about 1.5. The standard deviation is about 2.2. The words that appear more frequently than the average but not more than the average plus 1 standard deviation (to eliminate common words such as "and" and "of") are "we," "to," "structure," "protein," "for," "assessment," and "methods." We can easily eliminate the words "we," "to," and "for" on the basis of some dictionary of very common noninformative words or by taking into account the length of the words to obtain "structure protein assessment methods," which is not a bad summary, given the shortness of the abstract and the lack of sophistication of our algorithm.

Computational linguistics techniques can be quite effective. On the other hand, a domain-based grammar is very promising in the biosciences because of the nature of the text itself, which is less ambiguous than general text. The two approaches can, therefore, be combined, and such a combination is being applied extensively to the discovery of protein–protein relationships. Worldwide initiatives are underway to blindly evaluate the performance of these methods and provide an updated assessment of their reliability.

Transferring Functional Annotations by Similarity

What is the level of identity, or similarity, between two sequences that guarantees that they have the same function at some level of detail? If we had a precise answer to this question, a large number of proteins in our database could be automatically and correctly associated with a functional annotation. However, we do not have the answer, or, to be more precise, nature does not behave in such a simple way.

Sequence identity, or similarity, can, at most, guarantee the existence of an evolutionary relationship between two proteins. We have already discussed the problem of orthology and paralogy: if a gene has undergone both duplication and speciation, the existence of an evolutionary relationship does not guarantee an evolutionary pressure for maintaining a common function. We have also discussed another major problem: domain-boundary detection. If the evolutionary relationship is limited to a domain, transferring function might be seriously misleading. A schematic example of the disastrous effects such a transfer can have is shown in Figure 32.

For single-domain proteins, we can estimate our ability to detect function on the basis of evolutionary relationships by analyzing function conservation between pairs of known evolutionarily related proteins as a function of their evolutionary distance, which, for the sake of this discussion, we will approximate with their sequence identity after optimal alignment. Structural conservation, as we will discuss at length, is strictly related to homology, and, indeed, more than 80% of proteins that share at least 25% sequence identity have very similar structures. All pairs of proteins that share more than 45% sequence identity have very similar structure. The EC classification scheme allows us to analyze the extent of functional conservation between pairs of

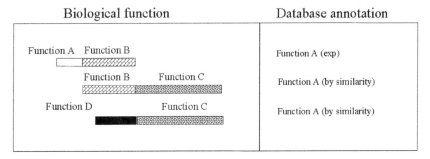

FIGURE 32

Schematic example of the danger associated with similarity-based annotation of proteins. A multidomain protein carries out two functions, A and B. Only function A is detected by the experimentalist, who annotates the entire protein as performing function A. Later a new sequence, sharing a similarity with the original protein, is incorrectly annotated as having function A. The error can be propagated to another sequence that shares essentially no similarity with the original experimentally annotated protein.

evolutionarily related enzymes, and here things become more problematic. Only proteins that share more than 85% sequence identity have strictly conserved function (up to the fourth digit of the classification scheme). The third digit is shared by proteins that have at least 55% sequence identity. At 25% sequence identity, not more than 60% of protein pairs have the same EC code, an additional 20% only share the first three digits of the classification, and a few percent of pairs have no common EC digit at all. If we compare SwissProt key words and, therefore, can also include nonenzymes, matters become even less hopeful. Not even 95% sequence identity guarantees strict conservation of annotation and, at a level of sequence identity of 25%, no more than 45% of the pairs have the same key words. Clearly, the numbers listed here are bound to change as methods and databases change, but they are useful as reference points.

A careful analysis of the alignment that takes into account *which* residues are conserved rather than how many are conserved can most likely improve matters quite a bit. However, in the genomic era, we cannot afford to assign function by carefully and manually analyzing every case. The world is waiting for results on tens of thousands of proteins, and our methods need to be fast and automatic.

Transcriptomics

The genome provides us with a static picture of the cell. It tells us which genes can be present, but it does not offer any information about which genes are translated into proteins in any given cell at any given time. Yet, this information is the crucial information: a liver cell and a brain cell contain the same genetic material, but they are very different morphologically and functionally because they express different proteins in different conditions. The knowledge of the genomic sequences of many organisms presents the possibility of also investigating these aspects by performing genome-wide experiments that can provide data, not on single genes and proteins, but on the whole set of genes and proteins of a cell. We can get a dynamic picture of what is happening in a cell at different times or in response to different stimuli or when a damaged cell is transforming itself, for example, into a cancer cell.

If we want to exploit the large body of information made available by the genomic efforts, we must grasp the experimental techniques that are being developed. Theoreticians must realize the importance of understanding the experimental methods and their limitations and reliability as much as experimentalists must have an idea of the underlying assumptions of the many computational tools at their disposal. For these reasons, we briefly outline the basics of transcriptomics and proteomics techniques without the experimental details. This simplification necessarily leads to imprecision, so we urge the readers who desire a more thorough understanding of these techniques not to rely solely on the explanations given here.

The idea behind transcriptomics experiments is the following: genes are copied (transcribed) into an RNA molecule before being translated into proteins, and the number of copies of the mRNA molecules roughly correlates with the amount of protein the cell will produce. This observation implies that a measure of the amount of mRNA coding for a given protein will tell us whether and approximately in what amount a protein is produced and whether perturbations of the system change its abundance.

However, the picture arising from the analysis of gene expression level can be very complex. A human has about 50,000 genes. Even if genes could only be on or off, the possible states would be $2^{50,000}$. However genes are not just switched on or off; expression levels can vary from 1 to 10^8 molecules per cell.

Assume that we have a sample of RNA molecules with a given sequence (probe) on a surface over which we deposit a complex mixture of RNA in which each molecule is labeled radioactively or with a fluorescent dye. If we wash away all molecules that do not bind (hybridize) to our surface, only the RNA molecules with a sequence complementary to our original molecule will be left, and, therefore, the presence of radioactivity or fluorescence will indicate their presence and the amount of radioactivity or the intensity of the color will reveal their concentration.

RNA is not a very stable molecule, whereas DNA is. Therefore, in these experiments, the extracted RNA is "copied" (retrotranscribed) into DNA by use of enzymes that catalyze this reaction. DNA obtained by retrotranscribing an RNA molecule is called cDNA. We can deposit on our surface, in different positions, different probe sequences and hybridize all of them at the same time with a mixture of cDNA extracted from a cell. The pattern of fluorescence will tell us which molecules (cDNAs and, therefore, with approximation, mRNAs) are present in our sample and at what concentration. This process is the general idea behind transcriptomic experiments.

We must deposit the probe molecules and obtain our sample. In one technique, a cDNA library is used: the total mRNA from a cell clone is collected, retrotranscribed into cDNA, and cloned into an appropriate host cell. The host cells are grown, and each of them will produce a clone that contains one of the original cDNA sequences. The DNA from each clone is extracted and "spotted" on the surface. The resulting "microarray" will contain, in each position, many cDNA molecules with the same sequence, complementary to a gene of the original cell.

A different technique (GeneChip® from Affymetrix) takes direct advantage of the knowledge of the sequence of the genes of an organism. Regions of DNA contained in each gene are chemically synthesized directly on the microarray. Usually each segment is 25 nucleotides long and 15 to 20 segments are selected for each gene. The desired molecules are synthesized directly on the surface of the microarray. For each segment of each gene, a

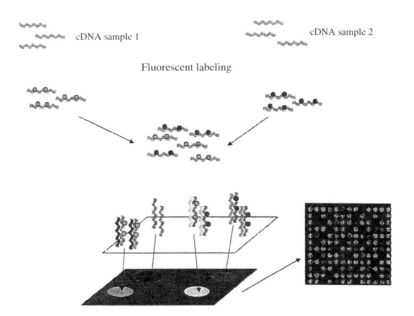

FIGURE 33

Scheme of transcriptomics experiments. cDNA molecules derived from different samples are treated with two different dyes, mixed, and made to hybridize with a previously prepared microarray. If the cDNA complementary to that contained in one of the microarray spots is present in both samples, the spot will be yellow. Red or green indicates that the molecule is only present in one of the two samples, and black indicates that it is absent in both.

sequence, differing from the original sequence by a single base change, is also synthesized and used as a control.

The RNA sample we want to investigate is extracted from the cells, retrotranscribed into cDNA, labeled with a dye, and hybridized to the microarray. The array is scanned with a laser, and the image is analyzed to extract numerical values for the dye intensities at each spot that allow us to estimate the presence and concentration of cDNA molecules in our sample (Figure 33) (see color insert after page 40).

If we have two samples (for example, a normal cell and a cancer cell), we can extract the RNA from both samples, retrotranscribe the RNAs, and label each RNA separately with a different dye. The two samples are subsequently mixed and hybridized to the microarray. If one sample is labeled with a red dye and the other sample is labeled with a green one, a yellow spot on the microarray will indicate equal amounts of the corresponding molecule in both samples, a reddish or a greenish spot will indicate overexpression of that molecule in one of the two samples, and a black spot will indicate that the specific RNA is not present in either cell.

The two techniques have advantages and disadvantages. In the cDNA library, we do not need to know the sequence of all the genes of the organism or cell we are analyzing. In the other technique (GeneChip®), we have a better statistical handle on the data because we have several spots for each gene and a negative (not supposed to hybridize) spot for each segment, namely the one with a mismatched nucleotide.

Both methods, however, are based on several important assumptions. RNA is not a very stable molecule and, therefore, we cannot be sure that all the molecules present in our sample will be retrotranscribed to the same extent. We must also assume that labeling is uniformly efficient in our molecules. Most importantly, we must assume that the amount of a protein in a cell is proportional to the amount of mRNA, a rather crude assumption because other mechanisms, such as the stability of the protein or of the mRNA itself, can regulate the concentration of the protein. Another disadvantage is that regulation of protein activities via posttranslational modification cannot be detected by these techniques.

Even with these limitations, these methods allow us to "see" what is happening to the transcription of genes at any moment in the life of a cell and to appreciate different transcriptional activities in different cells. This feature is a rather impressive tool for studying life at a system level rather than at an individual gene level.

Expression data can be used to understand not only which gene is expressed in a given condition but also whether different genes are activated in a coordinated manner. For example, genes that are all activated in response to a stimulus might cooperate in performing the required cellular function and, therefore, share some level of common functionality. We can analyze the expression of all genes of a single sample at different times or compare the variations in gene expression level between two differently treated samples. This use of expression data can be effective in detecting which genes are characteristic of a cell type with respect to others. For example, we can determine which genes are activated in a cancer cell with respect to a normal cell to understand what caused the cancer and what its molecular effects are, as well as to highlight specific proteins that are expressed solely in the cancer cell and target them for therapy. We can also use transciptomics data to detect patterns that are diagnostic of abnormal tissues. Furthermore, the cellular response to pharmacological treatments can be determined by observing the changes they induce in the expression patterns of the genes and correlating these to the efficacy of the treatment, thus presenting the possibility of personalized medicine.

The analysis of microarray data requires image analysis and background subtraction software, as well as statistical methods to obtain a biological interpretation of the data. We will not discuss the details of what can be done to eliminate the background noise and to normalize the spot intensity between different experiments. Although both are very serious problems, their solutions are common to many sciences and are nonbioinformatics-specific. We will focus on the problem of using the data to formulate sensible and testable biological hypotheses.

After a preliminary image and data analysis (i.e., background subtraction and intensity normalization), we can store the data in a matrix. Each row represents a gene, and each column represents a different experiment,

	Exp. 1	Exp. 2	Exp. 3	Exp. 4	Exp. 5
Gene 1	0.73	0.92	0.20	0.67	0.18
Gene 2	0.99	0.60	0.48	0.10	0.56
Gene 3	0.70	0.85	0.15	0.70	0.20
Gene 4	0.44	0.84	0.42	0.84	0.86
Gene 5	0.50	0.90	0.60	0.80	0.86
Gene 6	0.40	0.50	0.00	0.32	0.01
Gene 7	0.78	0.22	0.32	0.87	0.50
Gene 8	0.72	0.51	0.35	0.54	0.81
Gene 9	0.34	0.83	0.62	0.15	0.71
Gene 10	0.58	0.74	0.16	0.60	0.14
Gene 11	0.56	0.70	0.58	0.70	0.14
Gene 12	0.51	0.62	0.40	0.51	0.15
Gene 13	0.08	0.35	0.45	0.26	0.00

FIGURE 34
A data matrix for a microarray experiment.

timepoint, or cell type. In each cell of the matrix, we report, in arbitrary units, the relative expression level of the gene in the sample (Figure 34). This data matrix can be used to address several questions of interest, and we must use the appropriate tools to make the best use of our data matrix. We might ask several important questions: Which genes are differently expressed in different samples or conditions after different drug treatment? Which genes are responsible for a disease? Which gene expression levels can be used to characterize a given condition? Which genes belong to a network of activities that defines a general cell function, such as cell differentiation, or a particular metabolic pathway?

To detect which genes are similar (have similar behavior across samples) or which samples are similar (for example, which cell lines have the same set of genes highly expressed) we must define a measure of "distance" between the expression levels of genes, devise an algorithm for grouping (clustering) them according to their similarity, and evaluate the distance between groups (clusters) of genes.

A distance function d is called a semimetric measure if it satisfies the following requirements: $d(A,B)$ is always positive for $A \neq B$, equal to 0 for $A = B$, and independent on the order of the objects; that is, $d(A,B) = d(B,A)$. If the inequality $d(A,B) < d(B,C) + d(C,A)$ is also satisfied, then the distance function is called metric.

We can apply Euclidean distance to our expression measurements:

$$d(gene1, gene2) = \sqrt{(e_{gene1,1} - e_{gene2,1})^2 + \ldots + (e_{gene1,p} - e_{gene2,p})^2 + \ldots + (e_{gene1,n} - e_{gene2,n})^2}$$

where e_{ip} is the expression level of gene1 in the p-th experiment. This possibility is only one of the several available to us. We can use the Manhattan distance (a semimetric function):

$$d(gene1, gene2) = \left| e_{gene1,i1} - e_{gene2,1} \right| + \ldots + \left| e_{gene1,p} - e_{gene2,p} \right| + \ldots + \left| e_{gene1,n} - e_{gene2,n} \right|$$

We can also compute the Pearson correlation coefficient:

$$C(gene1, gene2) = \frac{\sum_i (e_{gene1,i} - \overline{e_{gene1,i}}) \cdot (e_{gene2,i} - \overline{e_{gene2,i}})}{\sqrt{\sum_i (e_{gene1,i} - \overline{e_{gene1,i}})^2 \cdot \overline{(e_{gene2,i} - e_{gene2,i})^2}}}$$

where $e_{gene1,i}$ is the average expression value for gene1 over all the experiments.

Other measures are available, and they are not equivalent, as we will illustrate with an example. In Figure 35, we show the Euclidean distance between pairs of genes on the basis of the data matrix in Figure 34.

We will focus on the first six genes, the expression levels of which are plotted in Figure 35. The gene with the shortest distance from gene 1 is gene 3,

	Gene 1	Gene 2	Gene 3	Gene 4	Gene 5	Gene 6	Gene 7	Gene 8	Gene 9	Gene 10	Gene 11	Gene 12	Gene 13
Gene 1													
Gene 2	0.72												
Gene 3	0.01	0.75											
Gene 4	0.63	1.00	0.60										
Gene 5	0.69	0.92	0.69	0.04									
Gene 6	0.48	0.94	0.42	1.29	1.48								
Gene 7	0.65	0.81	0.55	0.64	0.75	0.87							
Gene 8	0.60	0.35	0.55	0.28	0.33	0.91	0.29						
Gene 9	0.89	0.52	0.91	0.55	0.48	1.02	1.22	0.48					
Gene 10	0.06	0.71	0.04	0.67	0.78	0.21	0.53	0.55	0.80				
Gene 11	0.22	0.74	0.23	0.60	0.57	0.56	0.50	0.59	0.69	0.19			
Gene 12	0.20	0.58	0.19	0.67	0.71	0.24	0.49	0.50	0.57	0.08	0.08		
Gene 13	1.01	1.23	0.96	1.45	1.53	0.33	1.15	1.18	0.84	0.62	0.58	0.35	

FIGURE 35
Euclidean distance between the expression level of the 13 genes in Figure 34.

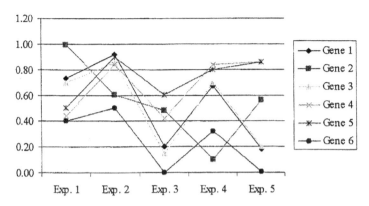

FIGURE 36
Graphical representation of the level of expression of the first six genes shown in Figure 34.

	Gene 1	Gene 2	Gene 3	Gene 4	Gene 5
Gene 1					
Gene 2	0.15				
Gene 3	0.99	0.09			
Gene 4	0.2	-0.5	0.26		
Gene 5	0.12	-0.5	0.15	0.95	
Gene 6	1	0.21	0.99	0.09	0.09

FIGURE 37
Pearson correlation coefficient for the expression level of the first six genes of Figure 34.

and the one with the longest distance is gene 2. This arrangement is intuitively correct, as seen in Figure 36 (see color insert after page 40).

However, gene 1 and gene 6 have a large Euclidian distance, but they
clearly behave in a correlated fashion. Their expression level increases in
Experiment 2 with respect to Experiment 1, decreases in Experiment 3,
increases in Experiment 4, and decreases again in Experiment 5 (Figure 34).
One gene might drive the transcription of the other. This relationship does
not necessarily imply a similar expression level, but, rather, causes a similar
trend in the modulation of their expression levels. If we use the Pearson
correlation coefficient (Figure 37), the correlated behavior of gene 1, gene 3,
and gene 6 becomes apparent.

Once we have selected the distance measure appropriate for the question
we want to ask, we can cluster our genes (or experiments) to highlight
similarities and reorganize our data matrix in a more informative way. The
final display has a permutation of the rows or columns of the matrix so that
adjacent rows or columns are similar. We can cluster our data by application

	Exp. 1	Exp. 2
Gene 1	0.73	0.92
Gene 2	0.99	0.60
Gene 3	0.70	0.85
Gene 4	0.44	0.84
Gene 5	0.50	0.90
Gene 6	0.40	0.50
Gene 7	0.78	0.22
Gene 8	0.72	0.51
Gene 9	0.34	0.83
Gene 10	0.58	0.74
Gene 11	0.56	0.70
Gene 12	0.51	0.62

FIGURE 38
A data matrix for 12 genes and two experiments.

	Gene 1	Gene 2	Gene 3	Gene 4	Gene 5	Gene 6	Gene 7	Gene 8	Gene 9	Gene 10	Gene 11
Gene 1											
Gene 2	0.41										
Gene 3	0.08	0.38									
Gene 4	0.30	0.60	0.26								
Gene 5	0.23	0.57	0.21	0.08							
Gene 6	0.53	0.60	0.46	0.34	0.41						
Gene 7	0.70	0.43	0.64	0.71	0.74	0.47					
Gene 8	0.41	0.28	0.34	0.43	0.45	0.32	0.30				
Gene 9	0.40	0.69	0.36	0.10	0.17	0.34	0.75	0.50			
Gene 10	0.23	0.43	0.16	0.18	0.18	0.30	0.55	0.26	0.26		
Gene 11	0.28	0.44	0.21	0.18	0.21	0.26	0.53	0.25	0.26	0.04	
Gene 12	0.37	0.48	0.30	0.23	0.28	0.16	0.48	0.24	0.27	0.14	0.09

FIGURE 39
Euclidean distance derived from the data matrix in Figure 38.

of a hierarchical strategy that employs the same methodology we described for the construction of sequence phylogenetic trees, as illustrated by the data matrix in Figure 38. The Euclidean distance matrix is shown in Figure 39 and plotted in Figure 40.

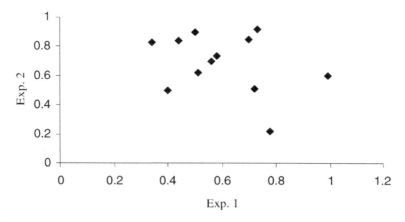

FIGURE 40
Plot of the data points of Figure 38.

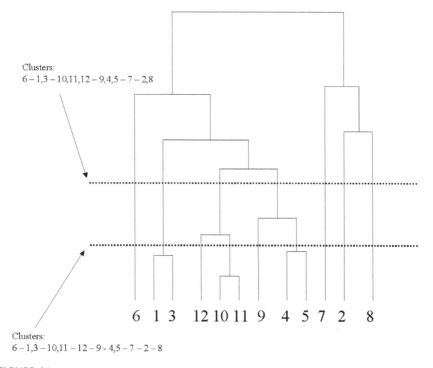

FIGURE 41
A tree representing the distances in Figure 39.

As we discussed in Problem 1, we can represent our data as a tree in which the length of the branches is proportional to the distance between the points. We can "trim" the tree at different levels. Each choice will provide us with a different number of clusters, as shown in Figure 41.

Another commonly used clustering method is the K-means clustering, in which the number of clusters is decided *a priori*. The procedure consists in randomly assigning each gene to a cluster and calculating the centroid of the cluster. The distances between the centroid and each gene are measured, and each gene is assigned (or reassigned) to the closest cluster. The procedure is repeated until convergence; that is, until each gene belongs to the closest cluster. A modification of the procedure is to change the assignment of each gene to a cluster randomly rather than on the basis of its distance from the clusters' centroids. The distance between different gene (or experiment) clusters can be defined as the distance between the closest or farthest pairs among all genes belonging to the two clusters, their average distance, or the distance between the centers of the clusters.

The grouping of our data and the corresponding reorganization of the rows and columns of the data matrix can highlight biologically relevant relationships between the data, but the results depend, among other things, on our choice of distance definition and cluster methodology. It is difficult to assess the reliability of the different analysis methods, as we do not yet have a large set of data with known properties. This problem is the major shortcoming in the analysis of correlation of microarray data.

Another problem is that the expression of each gene is often regulated by more than one protein, which might be active in different conditions. Distinct groups of genes may be coregulated in different samples, and, therefore, more complex methods are needed.

Whatever the analysis method, we end up with several sets of genes grouped together and deemed to have something in common. The next step is the evaluation of the biological significance of the results, which is especially difficult. Everything we learned about classification of functions should be applied here to determine whether genes clustered together by our data analysis are likely to share a function or to be part of a given pathway.

What we described above is useful if we want to detect relationships in our gene expression levels without any prior information. In some cases, we may have knowledge of the existence of groups of data and would like to know whether a new object (a gene or an experiment) belongs to one of our groups. For example, we might have collected expression data on normal and cancer cell lines and would like to know whether a new sample should be classified as derived from normal or cancer cells.

The problem becomes a classification problem, and we have already seen some techniques that can be used for this purpose, such as neural networks. The support vector machine (SVM) is another technique often used in this application. The idea is to represent our data as points in a multidimensional space and ask whether a hyperplane (i.e., a plane in a high dimensional space) exists that separates positive and negative examples with the largest margin.

The detection of diagnostic patterns in different microarrays has already found useful applications in clinical practice, such as classifying tumor types to create "fingerprints." However, the significance of the clusters in terms of diagnosis and prognosis of the tumor must be validated on a case-by-case basis. Similarly, gene expression patterns can be related to response to a given treatment, survival times, or activation of specific metabolic pathways.

Microarray data have a high dimensionality. Although we have a few measurements of the expression level of each gene under a limited set of conditions, we have thousands of genes to analyze at the same time. Thus, despite the advantages of high-throughput data collection, it presents a major challenge for data storage, visualization, and statistical interpretation. If we want to determine which of the thousands of genes change expression in two different conditions, the problem of multiple comparisons becomes quite severe.

In non-high-throughput experiments, we can measure the expression level of a single gene in different conditions several times and test whether the differences are statistically significant for a given confidence level. In array experiments, the number of repetitions is generally low. Furthermore, the chances of observing a spurious difference, which is nonbiologically significant, increase with the number of genes. The low number of repetitions and the large number of variables pose serious problems to the statistical analysis of microarray data.

The problem of the high dimensionality of the data is serious, and techniques to reduce the number of variables are continuously being developed. The adopted solutions range from excluding from the analysis genes that are known to have a quasiconstant level of expression across cells to the application of standard dimensionality reduction techniques, such as principal components analysis (PCA). PCA is a mathematical procedure that is used to transform a number of correlated variables into a smaller number of uncorrelated variables called principal components. The extraction of the first principal component corresponds to a rotation of the original variable space, such that the variance of the data is maximal when it is projected on the axis. The second principal component is the principal component of the residuals, and so on.

In a scatterplot, the first principal component is a new x-axis, rotated so that it approximates the regression line. An illustration of a simple case of PCA is shown in Figure 42.

Proteomics

Transcriptomics experiments allow us to have a dynamic picture of the level of transcription of each gene in a cell, but this picture is still insufficient to define the state of the cell. The mRNA is translated into proteins, and proteins have different life times, they interact with each other, and their activity is often regulated by posttranslational modifications, as we discussed in the case of glycogen phosphorylase.

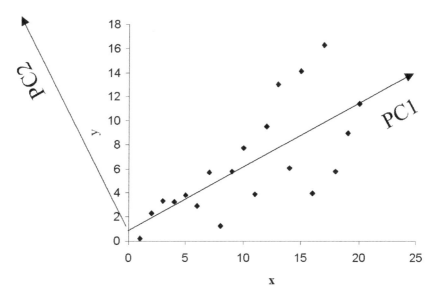

FIGURE 42
Principal component analysis for a set of artificially generated data. The projection on the first
principal component (indicated with PC1) is the one with maximum variance.

Analysis of the protein content of a cell, rather than the content of the
mRNA molecules that will be translated, is more informative. Unfortunately,
proteins are much less easy to identify than nucleic acids. We can deduce
the presence of a nucleotide sequence by its ability to bind (hybridization)
with a complementary probe, but we have no hybridization mechanism to
detect protein sequences.

The possibility of identifying and quantifying every protein in a cell was
unthinkable until a few years ago. The major stumbling blocks were sepa-
rating the large number of different proteins present in a cell at any given
time and, especially, identifying them. One way to separate different proteins
in a sample is to use a sodium dodecyl sulfate (SDS) gel. The charged SDS
molecules bind to a protein in a number proportional to the size of the
protein. If SDS-bound proteins are subjected to an electrical field, larger
proteins will have more charged bound molecules and, therefore, will move
faster than smaller proteins. At the end of the electrophoretic experiment,
the gel can be stained with dyes that specifically bind to proteins so that
proteins migrating at different speed, and, hence, with different mass, will
cause spots located at different distances from the start of the run. The
portion of the gel that contains the spot that corresponds to the protein of
interest can be (literally) cut from the gel to recover the protein. Depending
on the experimental setup, SDS gels can separate up to a few dozen proteins,
but not more than 100 in most cases. SDS gels are, therefore, unsuitable for
analyzing the full protein content of a cell.

Proteins can also be separated according to their isoelectric point (pI). The
pI of a protein is the value of pH at which a protein is electrically neutral.

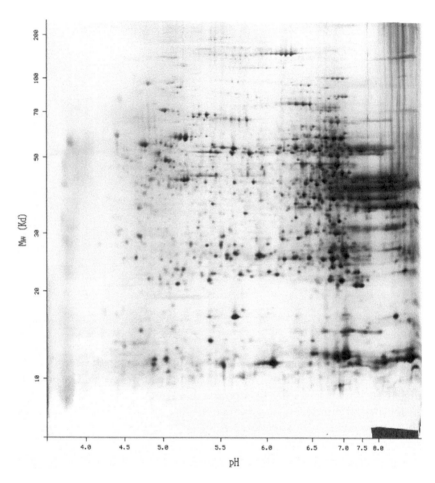

FIGURE 43

Two-dimensional gel for human liver (from the 2DWG meta-database of 2D-PAGE protein gel images: http://www.lecb.ncifcrf.gov/2dwgDB).

This value clearly depends upon the number of charged atoms in the protein and is different for different proteins. The protein can be run in a gel in which a stable pH gradient has been established and subjected to an electrical field. The protein will stop at the position in the gel where the pH value is identical to the pI of the protein (i.e., the protein will become neutral).

Isoelectric point and size are unrelated to each other. Therefore, these values can be used to separate protein in two different dimensions, as shown in Figure 43. These so-called bidimensional gels are able to separate up to 10,000 proteins, and, thus, can be used to investigate the protein content of a cell. Not every protein will be easily separated by two-dimensional electrophoresis. Membrane proteins, for example, are very hydrophobic and difficult to analyze. Similarly, proteins with very high pI (very acidic or hyperphosphorylated proteins) present a problem.

Moreover, direct quantification can be a problem because intensity of staining is linear, with concentration only over a very narrow range (0.04 to 2 ng/mm^2). A solution to this problem is not to rely on staining intensity for quantification, but on mass spectrometry. This technique is extremely sensitive and has a very high dynamic range. It can detect posttranslational modifications and can deal with different proteins in the same spot. Perhaps more importantly, it provides a way to unambiguously identify each protein in a sample.

Mass spectrometry detects and measures the number of ions with given mass-to-charge ratio, and it has traditionally been applied to volatile compounds. However, two relatively recent protein ionization techniques, matrix assisted laser desorbation/ionization (MALDI) and electrospray ionization (ESI), have revolutionized the field. In MALDI, a nonvolatile sample (e.g., a protein mixture) is embedded in a matrix and irradiated by a short laser pulse. The energy of the pulse is converted into heat that produces "evaporation" of the sample, which becomes ionized by a single proton. The ionized protein is then analyzed by mass spectrometry. In ESI, the sample is embedded in a ionized solvent drop, electrospray causes concentration of the solvent in the drop by evaporation and, thereby, increases the charge of the drop. The sample in the drop becomes highly charged, and its mass-to-charge ratio can be measured by mass spectrometry. Techniques to separate proteins before analysis can be used in conjunction with mass spectrometry.

How we can identify which protein corresponds to each spectrometry peak? Clearly, the molecular mass is not sufficient. Many proteins have the same molecular mass, and, therefore, we need to add further steps to our procedure. For example, we can let our sample pass through a collision chamber where it is broken into pieces, and we can use the mass-to-charge ratio of each peak as additional information to identify our protein. We can also specifically cleave our sample by use of a protease. Several enzymes that hydrolyze peptide bonds adjacent to specific amino acids are known, and the characterization of the mass of each fragment, together with the knowledge of the type of amino acid that is adjacent to the cleavage site, can aid in the search of our database of protein sequences for a protein likely to be the one under analysis.

When a database search is based on a short sequence fragment as query, use of more than one fragment from the same protein is essential to avoid misassignment of proteins to peaks. We can also use mass spectrometry to obtain the actual amino acid sequence of our fragments.

Promising Avenues

We conclude this problem by noting that, although here we surveyed different approaches to the function detection problem, from homology to data extraction from the literature, from microarray data to proteomics experi-

ment, all of the approaches would greatly benefit if a good standard definition of function were agreed upon and rationalized, if different data types
were synergistically combined in interpreting the results, and if a sound
treatment of error propagation in databases were developed. For example,
although GO ontologies allow us to annotate genes and their products, they
do not include annotation related to which cells or tissues express the genes
and their products, at which developmental stages they are functional, or
whether they are involved in disease. Other ontologies are devoted to these
aspects. The challenge is to create tools that maximize the utility of each
ontology while avoiding redundancy. Analogously, functional annotations
derived from databases such as SwissProt or from the literature itself could
be used as starting data together with the experimental data in clustering
techniques for genomics experiments.

The keywords here are integration and error minimization. Integration
should be achieved with scalable and user-friendly tools. It cannot be done
independently by several groups in different areas, but should rely upon
standards devised from consortia of experts. GO consortium and other ontologies are good models for this development.

A very serious problem is error propagation. A network of correlations
and interactions is behind our data. Each protein interacts with many others,
is controlled by many others, and, more importantly, has been assigned an
annotation mostly on the basis of its evolutionary relationships with other
molecules. What happens if one of our annotations is wrong? How does this
propagate to other elements of our network?

As in many areas, the network of annotations in the database, that is, the
network that describes which protein function has been annotated on the
basis of which other function, seems to be a "scale-free" network. Scale-free
networks are those in which the distribution of the connections is very
uneven. Some nodes have many more connections than others. In general,
scale-free networks can be described by a simple relationship: the probability
$P(k)$ that a node has k connections is proportional to $k^{-\lambda}$. Scale-free networks
describe several systems, from collaboration in scientific papers, to protein
interactions, to the World Wide Web (for which the value of λ is about 2.2).
Scale-free networks are safer than randomly connected networks. Random
network failures are more likely to happen in less connected nodes because
more of them exist, but the highly connected nodes are very sensitive spots
of the whole system.

How does an error propagate in a scale-free network? In other words, if
one computer in the network fails, how many other computers are affected?
The answer depends upon which site goes down. If the site is a node with
many connections, the error will propagate quickly and the effect can be
very relevant. The same is the case for protein functional annotations: the
wrong annotation of a very popular protein (i.e., a protein belonging to a
large evolutionary family) will produce much more damage to the consistency of the data than an error in the annotation of a protein only present
in a very specialized cell type or one organism. A formal treatment of error

propagation in scale-free networks is not available as of today, yet it would be invaluable for allowing a more controlled and effective verification of the level of errors in our databases.

Suggested Reading

Boeckmann B., Bairoch A., Apweiler R., Blatter M.-C., Estreicher A., Gasteiger E., Martin M.J., Michoud K., O'Donovan C., Phan I., Pilbout S., Schneider M. *The SWISS-PROT protein knowledgebase and its supplement TrEMBL in 2003, Nucleic Acids Res.* 31, 365–370, 2003.

The Gene Ontology Consortium. Gene ontology: tool for the unification of biology, *Nat. Genet.* 25, 25–29, 2000.

Quackenbush, J. Computational analysis of microarray data, *Nat. Rev.* 2, 418, 2001.

Celis, J. E. and Gromov, P. 2D protein electrophoresis: can it be perfected? *Curr. Opin. Biotechnol.* 10, 16–21, 1999.

Karas, M., Bachmann, D., Bahr, U., and Hillenkamp, F. Matrix-assisted ultraviolet laser desorption of non-volatile compounds, *Int. J. Mass Spectrom. Ion Process.* 78, 53, 1987.

Burlingame, A. L., Carr, S. A., and Baldwin, M. A. *Mass Spectrometry in Biology and Medicine*, Humana Press Totowa, NJ, 2000.

Covey, T. Liquid chromatography/mass spectrometry for the analysis of protein digests, *Methods Mol. Biol.* 61, 83–99, 1996.

Kinoshita, K. and Nakamura, H. Protein informatics towards function identification, *Curr. Opin. Struct. Biol.* 13, 396–400, 2003.

Andrade, M. A. and Valencia, A., Automatic annotation for biological sequences by extraction of keywords from MEDLINE abstracts, *ISMB97*, 5, 25–32, 1997.

Problem 4

Protein Structure Prediction

Introduction to the Problem

The biochemical function of a protein is determined largely by its three-dimensional structure. In turn, the structure of a protein is mainly dictated by the specific linear sequence of its amino acids, as first demonstrated by Anfinsen in a historical experiment. He showed that a protein (in that particular case, ribonuclease A) once denatured—that is, unfolded—*in vitro* recovers its "native" conformation when the denaturing agents are removed from the test tube. The same experiment can be conducted with a chemically synthesized protein, which implies that the information about the three-dimensional structure of a protein is contained in its amino acid sequence. Subsequently, cellular mechanisms were discovered that catalyze folding of some proteins. These systems accelerate the folding process, but do not affect the structure of the final native state.

The structure of a protein can be experimentally determined by use of techniques such as X-ray crystallography and nuclear magnetic resonance, but these techniques are time and labor consuming, and not all the proteins of the universe can be experimentally determined. Therefore, we would like to infer the three-dimensional structure of a protein, given its amino acid sequence.

Energetic Calculations of Protein Structures

Energy Calculation

The observation that a protein, in appropriate conditions, folds into the same stable structure implies that this native structure is the global free-energy minimum among the states that the proteins can explore. The existence of a free-energy minimum is not guaranteed by general physical properties of polymers, but by evolution. In other words, almost every native protein sequence has been selected by evolution to fold into a single, stable structure.

A protein achieves a stable structure if the energy lost upon folding is compensated by the energy gained in the folded state. During folding, a protein loses entropy because entropy is related to the number of states available to a system, and an unfolded chain has a practically infinite number of possible conformations. The protein chain also loses the energy associated with the hydrogen bonds that its polar atoms form with the solvent.

Upon folding, a protein gains both entropy and enthalpy. Entropy is gained because of the burial of hydrophobic side chains of amino acids. When these side chains are exposed to a polar solvent in the unfolded state, they cause ordering (i.e., loss of entropy) of the surrounding polar solvent, and this condition is energetically unfavorable. This effect explains the tendency of nonpolar solutes, such as oil, to form drops in polar solvents, such as water. The sphere has the minimum surface for a given volume, and, therefore, a sphere causes the ordering of the lowest possible number of polar atoms. In a protein, apolar groups are shielded from the solvent and buried within the core of the molecule, which results in an energetic gain as the molecule reduces the loss of entropy with respect to the unfolded state of the protein. A protein gains enthalpy because of the many, generally weak, interactions that are established in a protein structure. These interactions include intra-chain hydrogen bonding, as well as VanderWaals and electrostatic interactions. A protein can assume a unique energetically favorable structure because it contains both polar and hydrophobic atoms. The latter make unfavorable interactions with a polar solvent and must be buried in an apolar environment. (Membrane proteins, that is, proteins embedded in apolar membranes, are discussed separately.)

Proteins are only marginally stable. Therefore, although the free-energy values of the unfolded state and of the native state are rather high, the difference between them is very small. To estimate the free-energy values with a reasonable accuracy, we must estimate the energy of the folded and unfolded state with a level of precision that is not currently achievable. Therefore, calculating the energy of every possible conformation of a protein chain to select the native low-energy state is impossible not only because of the enormous number of conformations but also because of the more fundamental problem of the lack of precision of our calculations. However, methods are available by which to obtain approximate estimates of the energy associated with a given protein conformation, and although they are too inaccurate to allow us to predict the structure of a protein *a priori*, they are useful in many applications.

The problem of exploring the conformational space is equally complex. As pointed out by Cyrus Levinthal in 1969, the conformational space available to a protein is enormous. We not only lack the ability to evaluate the energy of each of the possible conformations but also face the problem of understanding how nature does it. Statistically, the average time of the folding process should be comparable to the time needed to try every possible conformation. A simple calculation shows that if this condition were applicable, the lifetime of the universe would not have been sufficiently long to allow even a single protein to fold. Recent theories propose possible solutions to this problem, which is usually referred to as the Levinthal paradox. We are, thus, beginning to understand, at least in principle, how a protein can achieve its structure in a matter of milliseconds to seconds.

Molecular Mechanics

The interactions between protein atoms are governed by quantum mechanics. A precise treatment of the problem is currently impossible and will remain so for some time. Therefore, a classical mechanical approximation must be applied. For example, two covalently bound atoms can be seen as two classical particles, with a mass, linked by a spring whose elastic constant is proportional to the strength of the covalent bond. According to Hooke's law,

$$F_r = K_r(r - r_0)$$

The equilibrium distance r_0 between every pair of atom types (for example, two carbon atoms linked by a double bond) can be derived by a statistical analysis of the distances observed between the two atoms in experimentally determined protein structures or by precise quantum-mechanical calculations on smaller, and, therefore, more tractable, model systems. The same holds for the set of parameters K_r.

A similar procedure can be applied to estimate the energy associated with other bonded interactions, such as the angle θ between each possible triplet of atoms or the dihedral angles ϕ between each possible quadruplet. Commonly used approximations are

$$K(\theta - \theta_0)^2 \text{ and } \tfrac{1}{2} K_\phi(1 + \cos n\phi)$$

in which the same strategies outlined for the covalent bond interactions are used to derive the parameters.

Nonbonded interactions are more difficult to model. A widely used approximation for VanderWaals interactions is

$$A_{ij} r_{ij}^{-12} - B_{ij} r_{ij}^{-6} \tag{1}$$

where r_{ij} is the distance between the two atoms, and A_{ij} and B_{ij} are again empirically parameterized. Incidentally, the VanderWaals interaction is better modeled by different functions; the one used was historically selected because it could be easily computed.

A major problem is the treatment of charge–charge interactions. We can apply the Coulomb law, but we must take into account all charged and partially charged atoms, including the solvent, and we must model multi-charge systems, which is very difficult.

Hydrogen bonds are electrostatic interactions and, therefore, should not require a separate treatment, but the approximations introduced in our calculation of charge interactions makes introduction of an *ad hoc* term to take them into account, which is, in some cases, convenient:

$$C_{ij} r_{ij}^{-12} - D_{ij} r_{ij}^{-10} \tag{2}$$

In general, the formula we use to calculate the energy of a given arrangements of atoms in a protein is of the following type:

$$E = \sum_{bonds} K_r (r - r_0) + \sum_{angles} K((--_0))^2 + \sum_{\substack{dihedral \\ angles}} \int K_\phi (1 - + \cos n\phi))$$

$$+ \sum_i \sum_{j<i} (A_{ij} R_{ij}^{-12} - B_{ij} R_{ij}^6) + \sum_i \sum_{j<i} (C_{ij} R_{ij}^{-12} - D_{ij} R_{ij}^{-10}) \tag{3}$$

The parameters in Equation (3) are collectively called "force field," and different sets are publicly available.

If Equation (3) were sufficiently accurate to correctly evaluate the energy of a given protein structure, we could search for the protein conformation that minimizes the value of the energy and this conformation should correspond to the native protein structure. As we said, the computation of the energy is not precise because of the unavoidable approximations introduced by our classical treatment of a quanto-mechanical problem. The energy values derived by equations of this type can only be used to distinguish between correct and incorrect protein arrangements when the incorrect arrangement is very different from the correct one. These values are, in general, unable to consistently distinguish between a correctly folded and an incorrectly folded protein if the latter is not very different from the former.

Equation (3) can be used in combination with optimization techniques to search the conformational space more efficiently. Algorithms such as Monte Carlo simulations, simulated annealing, molecular dynamics, and genetic algorithms have been tried and also applied in combination with different methods for evaluating the energy and with other heuristic methods.

Potentials of Mean Force

The Boltzmann equation states that the probability of a microstate with energy E in equilibrium with a thermal bath at temperature T is proportional to $e^{-E_i/KT}$. The probability of the various states must add to 1. Therefore, the probability of a microstate *i* with energy E_i is

$$p_i = \frac{e^{-E_i/KT}}{\sum_i (e^{-E_i/KT})} = \frac{e^{-E_i/KT}}{Z} \tag{4}$$

where Z is called the partition function. In simpler terms, states with higher energy are more unlikely to be populated. The equation can be inverted:

$$E_i = -KT \ln(p_i) + KT \ln Z \tag{5}$$

Z cannot be computed, but we can get rid of it if we compute the energy difference with respect to a reference state:

$$\Delta E_i = E_i - E_{ref} = KT \ln \frac{p_i}{p_{ref}} \tag{6}$$

We can deduce the energy of a state from its occupancy, provided we know the occupancy of the reference or ground state.

The Boltzmann equation holds when fluctuations occur between the states; that is, when the particles can move freely from one state to the other. This allows the particles to distribute themselves in such a way that the probability that a state i is occupied is proportional to $\exp(\Delta E_i/KT)$, and ΔE_i is the difference in free energy between the state i and the ground state.

The energy and the probability of occurrence of a given interaction in a protein structure are also linked by the inverse Boltzmann equation; that is, the probability of occurrence of an interaction between two amino acid side chains at a given distance can be used to calculate the strength of their interaction. In other words, very common interactions in protein structures are energetically more favorable than rarely observed interactions. This observation might seem surprising, because amino acids are not free to fluctuate from one state to another (i.e., to change their position in the protein). However, the number of sequences able to stabilize a given fold decreases with the energy of the fold, and, therefore, the most frequently observed sequence combinations are likely to be the most energetically favorable.

Because we know a large number of protein structures, we can estimate the probability of an interaction from its frequency of occurrence in known protein structures and, consequently, derive its energy. We can compute the ratio between the observed frequency of an interaction in the large collection of available protein structures and the frequency with which that interaction is expected to occur by chance alone. This method can be used to estimate the energy of the interactions that occur in a given protein conformation.

This approach has several advantages, such as the speed at which the energy can be calculated and its relative effectiveness, but it also has a number of complications.

First, protein atoms are not independent, because they are covalently bound to each other (e.g., two linked amino acids are close to each other because they cannot be otherwise). The frequency distribution for each pair of amino acids must be computed for different sequence separations (e.g., two alanine residues separated by three, four, five, or more intervening amino acids in the protein sequence). Second, we must answer the following questions: What do we compare our distribution with? What is the expected distribution of the distance between pairs of alanine residues separated by the given number of intervening amino acids in the unfolded protein structure? Which structure is the unfolded conformation? These questions are not trivial. The unfolded state cannot be easily studied, because it is an ensemble of states, and our experimental methods cannot be used to study the details of the conformations. We can use decoy structures; that is, build a computational model of the ground state. The choice of the ground state has an effect on our results. For example, if we use randomly generated amino acid chains, they might not contain secondary structure elements, and we will note that our distribution of distances in real protein structures will have a maximum at distances of around 5 Å. This distance corresponds to two residues in subsequent turns of α-helices, a structure that is observed more frequently in a protein structure than in an ensemble of random polymers. If we use shuffled protein structures as our ground state (i.e., if we randomly exchange the position of amino acids in real protein structures), our random distribution will contain a sizeable number of amino acids in α-helices, and, therefore, the peak at 5 Å will be above the background only for pairs of amino acids that are more frequently in helices than expected by chance alone. Third, we limit ourselves to pairwise interactions, which are a rather crude approximation of physical reality. Fortunately, pairwise potentials have proved to be extremely useful in several applications, and they appear able to capture general structure-related properties of a protein sequence.

Searching the Protein Conformational Space

Molecular Dynamics

Given an energy function, be it a statistical potential or a molecular mechanics one, we must search the conformational space of a protein to find the minimum of our energy function. A simple minimization algorithm changes the coordinates of our atoms, and, generally, a new conformation is accepted only if its energy is lower than the previous one. Therefore, it finds the local minimum closer to the initial conformation. Our global energy minimum

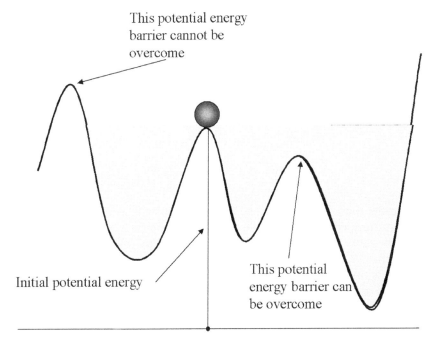

FIGURE 44
A sphere on a surface.

might be separated from our starting position by a potential barrier, and we will never explore it.

One solution is to endow our system with some initial energy that can allow potential barriers to be overcome. In mechanical terms, a ball deposited on a surface reaches the closer local minimum and overcomes local barriers if they are lower than the initial potential energy of the ball (Figure 44). In the absence of dissipative forces (i.e., with no friction in the example), the ball explores the conformation space confined between barriers with energy equal to its initial energy (the shaded area in the Figure 44). The higher the initial energy, the higher the potential barriers that can be overcome and, therefore, statistically, the larger the conformational space that can be explored. This idea is the basic concept of molecular dynamics.

If each of the atoms of our protein is provided with some "reasonable" kinetic energy, we can increase the size of the conformational space that can be explored. The velocity of the atoms is determined by a random sampling of velocities from a Maxwell-Boltzmann distribution that corresponds to a desired target temperature T. The fraction $f(v)$ of atoms having a specific velocity v is

$$f(v) = \frac{4}{\sqrt{\pi}} \left(\frac{m}{2k_B T} \right)^{3/2} v^2 \exp\left(\frac{-mv^2}{2K_b T} \right) \qquad (7)$$

where K_b is the Boltzmann constant, m is the mass of the particle, and T is the absolute temperature.

We can select an initial temperature and attribute an initial velocity (i.e., kinetic energy) to each atom in such a way that the overall distribution of velocities follows the Maxwell-Boltzmann distribution: the higher the temperature, the higher the proportion of atoms with a large velocity.

In this hypothesis, the atoms of our protein follow, in each instant of time, the Newton equation:

$$\vec{x} = \vec{x}_0 + \vec{v}_0 \Delta t + \frac{1}{2} \vec{a} (\Delta t)^2 \qquad (8)$$

We know the initial position of the protein (the coordinates of the atoms in our starting conformation), its velocity at time 0, calculated according to the Boltzmann-Maxwell equation and dependent upon our choice of the temperature T, and its instant acceleration, given by $\vec{a} = \vec{F}/m$ where $\vec{F}$ is the force acting on the atom that can be calculated from its potential energy calculated according to Equation (3). The subsequent position of each atom can be calculated by integrating the Newton equation, provided that the acceleration remains constant during the integration step; that is, provided that the integration step is very small. The requirement of a very small integration step limits the total time of our simulation, and, therefore, molecular dynamics can be used for exploring conformations reasonably close to the initial one, but molecular dynamics calculations are not very useful for exploring the large conformational space of a protein.

Monte Carlo Methods

Stochastic methods for the exploration of conformational space do not try to simulate a physical process, but rather to randomly explore the conformational space. The search is biased towards areas where the global minimum is more likely to be found. Monte Carlo methods use random moves; for example, one angle in the protein structure is randomly changed and the energy of the resulting conformation is calculated. Applied to our example of the ball, this method makes the ball disappear from its initial position and reappear randomly in a different position. If the new state has a lower energy than the previous one, we keep the move. However, conformations with higher energy are not necessarily discarded. They can be accepted with a probability dependent upon a preselected parameter, called temperature, and upon the difference between their energy and the energy of the previous state. The lower the increase in energy is, the higher the probability of accepting the move (Figure 45). In practice:

$$P = \min(1, \exp\frac{-(E_{new} - E_{previous})}{kT}) \qquad (9)$$

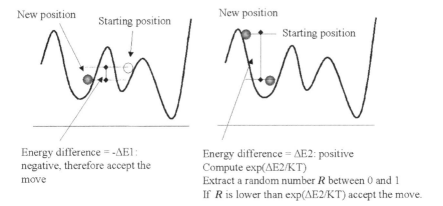

Energy difference = -ΔE1: negative, therefore accept the move

Energy difference = ΔE2: positive
Compute exp(ΔE2/KT)
Extract a random number R between 0 and 1
If R is lower than exp(ΔE2/KT) accept the move.

FIGURE 45
Scheme of the Metropolis Monte Carlo procedure.

If the new conformation has an energy E_{new} lower than the energy of the starting state $E_{previous}$, then $-(E_{new} - E_{previous})$ is positive, and the exponential is higher than 1. Therefore, the probability that the new conformation will be accepted is 1 (i.e., we always accept it). In the opposite case, the exponential is less than 1. The higher the difference in energy is, the lower the value of the exponential, and the lower the probability of accepting the new conformation.

Simulated Annealing

Both in molecular dynamics and in Monte Carlo methods, the parameter T does not necessarily represent a physical temperature. In both techniques, higher T values allow a wider exploration of the conformational space; lower T values ensure that the explored conformations have reasonably low energy. We, therefore, can start with a high temperature to allow the atoms to overcome larger barriers in molecular dynamics or to allow states with higher energy to be used for exploring new conformations in Monte Carlo methods. We then slowly decrease the temperature as the simulation proceeds, to progressively focus on areas with lower energy. The protocol used for decreasing the temperature (linear, stepwise, or a combination of both) depends on the specific method, but all are known under the name of simulated annealing.

Genetic Algorithms

Genetic algorithms are optimization algorithms based on the concepts of evolution. They start from an ensemble of possible solutions, such as a set of conformations of a protein structure. Similarly to what happens in molecular evolution, each element of the ensemble has a certain probability of being mutated (for example, by randomly changing a dihedral angle) or "crossed" with another element of the ensemble (for example, joining the first part of one conformation with the second of another, and vice versa). A fitness value is evaluated for each of the members of the population, and a new ensemble (generation) is constructed. The distribution of possible solutions in this new ensemble depends upon the fitness of each member. Members with higher fitness will be more represented than those with lower fitness, and the process will be repeated. The fitness function depends on what we want to optimize. In the case of protein structures, it can be the molecular mechanics energy of each conformation or, much more commonly, the mean field potential.

Knowledge-Based Methods

None of the methods described above is able to deduce the three-dimensional structure of a protein from its sequence, both because of the intrinsic limitations of our approximate energy calculation and because our ability to explore the conformational space is currently inadequate. However, the large set of available protein structures gives us a way to find a solution to the problem on the basis of heuristic rules that we can learn from the set of available solved examples (i.e., of proteins for which both the sequence and the structure are known). Although, even if we could use these heuristic methods to predict with satisfactory accuracy the structure of every protein in the universe, they would not necessarily help us understand the nature of the folding process. Nevertheless, we must exploit every possible route to find answers to our questions.

Evolution-Based Methods

The structure of a protein is determined by its amino acid sequence, and, therefore, the substitution of one amino acid with another affects the structure. In principle, the effect can be that the protein no longer folds (as is too often observed in random mutagenesis experiments in the laboratory), because the newly introduced interactions shift the energy balance toward the unfolded rather than the folded state. Alternatively, the substitution might not destabilize the protein structure and either be accommodated into the structure with only local minor rearrangements or cause the protein to assume a completely new unrelated structure. Because a protein structure is only marginally stable and this stabilization is mediated by a large number of weak interactions, and because the number of "foldable" sequences is a small set of all possible protein sequences, the last possibility is extremely unlikely and, indeed, has never observed. Therefore, a mutation either destabilizes a protein or is accommodated in a structure quite similar to the original one.

Evolution provides us with many examples of proteins that descend from a common ancestral protein whose sequence has been modified via a process of residue substitutions or, albeit less frequently, of insertions and deletions of amino acids. We know that these proteins are functional, or at least not deleterious, because they have been accepted in the population. Therefore, they are expected to have a stable native structure and also a similar structure. Clearly as mutations accumulate, local rearrangements also accumulate, so that the longer the evolutionary distance between the two proteins, the less conserved their structures.

As discussed in Problem 1, we have methods to detect evolutionarily related proteins from their sequences, and this ability implies that we can detect proteins very likely to have similar structures. The structure of one protein, therefore, can represent an approximate model for the structure of all the proteins of its evolutionary family. The closer the evolutionary distance is, the better the approximation. This concept is the basis of the most used method for protein structure prediction: homology, or comparative modeling.

The procedure is conceptually simple. If we assume that the sequence alignment between two protein sequences, one of unknown (the target) and one of known (the template) structure, obtained by one of the methods described in Problem 1, reflects the evolutionary relationship between their amino acids, we can assume that most of them have preserved the same relative position in the structure and use the coordinates of the backbone of the template as first approximations of the coordinates of the backbone of the target. We must model the conformation of the side chains and the local rearrangements of the structure brought about by the amino acid substitutions.

The assumption that we can find a correspondence between each amino acid of our target protein with one of the template is an oversimplification. First, if sequence changes have accumulated in less constrained parts of the protein structure (i.e., at its periphery rather than in the closely packed core), they might have produced substantial local rearrangements of the chain that must be modeled. Second, no region of the template corresponds to inserted amino acids, and the regions that surround deletions have necessarily changed their conformation to accommodate the new local sequence. These observations imply that we must identify the regions that can be modeled directly from the template (the conserved core) and, subsequently, model the remaining parts of the structure without the possibility of taking advantage of the template structure.

Templates for a comparative model can be found by searching the database of known protein structures for proteins putatively homologous to the target protein. If we find more than one, we have several possibilities. We can use only the template with the highest similarity to our target protein; we can select different templates for different regions of the target, taking into account the local sequence similarity; or we can "average" the structures of the templates. All these approaches are commonly used with quite similar results.

The orthologous/paralogous classification is less relevant here because the considerations that we made about structure conservation do not rely on the assumption that the evolutionarily related proteins perform the same function. Conversely, the problem of identifying the correct alignment is extremely relevant because an incorrect alignment can have disastrous consequences on the whole model. All that we said in Problem 1 about the importance of using reliable methods and multiple sequences is at least as important here as in function assignment.

As far as the conserved core is concerned, the quality of our model depends essentially on how correctly we have aligned the two sequences and on how much the two structures have diverged, or, in other words, on the extent of the local rearrangements required for accommodating replaced amino acids. We can easily learn how much the structures have diverged by an analysis of known protein structures, as demonstrated in a pioneering work by Lesk and Chothia. We can select pairs of evolutionarily related proteins, optimally superimpose them, and measure their structural divergence as a function of their evolutionary distance, estimated from their sequence difference. We define the structural divergence between two protein structures as their root mean square deviation (rmsd) defined as

$$rmsd = \sqrt{\frac{1}{N} \sum_{i=1}^{N} \left[(x_i - x_i')^2 + (y_i - y_i')^2 + (z_i - z_i')^2 \right]} \qquad (10)$$

where (x_i, y_i, z_i) and (x_i', y_i', z_i') are the coordinates of corresponding atoms.

Figure 46 shows the rmsd of the backbone of the core of pairs of evolutionarily related proteins as a function of the percent of identity between their amino acid sequences. The definition of the "core" of the structure differs in different methods. It can be intuitively seen as the internal, closely packed, conserved part of the structure that contains most of the repetitive secondary structure elements.

We can use the plot in Figure 46, for example, to estimate that, given a target protein and a model built from a template that shares 50% sequence identity with it, *if* we are able to correctly align the sequences of the two proteins, the average error that we expect for the coordinates of the backbone atoms of the core of our model is around 1 Å. Modeling of the regions that do not belong to the common core of the proteins is much more difficult because they can be structurally divergent as a result of local rearrangements, insertions, or deletions.

Regions in which local rearrangements are likely to have occurred can be recognized by their low local sequence similarity with the template and by

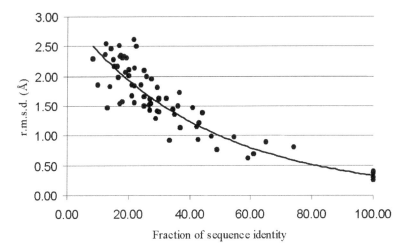

FIGURE 46
The relationship between sequence difference and structural divergence. The plot shows the value of the rmsd for the superposition of the core of seventy pairs of homologous proteins as a function of their sequence identity.

their localization near the surface of the template protein. Often, they can also be identified by use of different methods of sequence alignment and recognizing which regions of the alignment are less stable; that is, regions for which the alignment changes when we change the alignment parameters even slightly (indel penalties and similarity matrices)

Modeling of these regions is a continuing problem, although several methods have been explored. One method is to use energy-based strategies to predict structure, such as building all possible stereochemically reasonable conformations of the target protein backbone and evaluating their energy. Methods recently developed for proteins not sharing any sequence or structural similarities with known proteins have also been used (see below).

A very commonly used method is to search the protein structure database for regions that might constitute good local templates for the loop regions. Usually, these methods search for loops of the appropriate length in known proteins, the flanking regions of which fit well in the regions before and after the loop to be modeled. In some cases, these methods provide a reasonably good answer, but estimating their accuracy *a priori* is quite difficult. One important exception to the uncertain outcome of a loop prediction is seen in the case of immunoglobulin antigen-binding loops, as we discuss in Problem 7.

Once the backbone has been modeled, we must build the conformation of the side chains. We can model them by use of a packing or energy-optimization procedure, or we can use the known examples available, for example, by positioning each side chain in the conformation most often observed in the database of protein structures. Given the experimental backbone structure, several methods can reconstruct the side chains correctly, and, therefore, any improvement in modeling the backbone allows us to obtain reasonable results for the modeling of side chains.

One very successful optimization technique for side-chain conformation is an application of the dead-end elimination theorem, which is based on the assumption that the protein consists of a fixed backbone and a set of interacting side chains that can only assume a number of discrete conformations (rotamers). Given the backbone structure, the potential energy E_{total} of a protein conformation can be written as

$$E_{total} = E_{backbone} + \sum_i E(i_r) + \sum_i \sum_j E(i_r, j_s) \qquad i < j \qquad (11)$$

where $E_{backbone}$ is the energy of the backbone template, $E(i_r)$ is the energy of side chain i in the conformation r, and $E(i_r j_s)$ is the interaction energy between the amino acids i and j with rotamers r and s, respectively.

The dead-end elimination theorem states that some values of the rotamers are incompatible with the global energy minimum conformation, and these values are the rotamers i_r such that

$$E(i_r) + \sum_j \min_s E(i_r j_s) > E(i_t) + \sum_j \max_s E(i_t j_s); \quad i \neq j \qquad (12)$$

In other words, a rotamer can be excluded if its most favorable interactions have higher energy than the sum of the less favorable pairwise interactions that the same amino acid can make in any alternative conformation. In real cases, excluding these rotamers reduces the search space for side-chain conformations by many orders of magnitude.

At the end of the modeling procedure, we have an approximate structure for our protein, mainly obtained by use of the coordinates of other, evolutionarily related but not identical, proteins, and, ideally, we want to optimize the structure, taking into account the actual sequence of our target protein.

Unfortunately, none of the available energy-based optimization methods improves upon the starting model. If a method were able to optimize the starting model, then, given the model, the template structure used to build it, and the subsequently determined target structure, the model should be closer to the real structure than to the template used to build it. So far, no method has convincingly achieved this result because no modification of the template backbone structure has succeeded in obtaining coordinates significantly and consistently closer to the experimentally determined protein structure.

A recently developed technique is to generate, for each target protein, more than one model; for example, by use of different templates, by use of suboptimal as well as optimal alignments, and by use of diverse methods for predicting structurally divergent regions and for modeling the side chains.

After a large set of models has been obtained, they are ranked according to some fitness function (e.g., pairwise potentials) to select the putatively best one. These approaches appear more promising than the construction of a single model obtained by separately optimizing every step of the procedure.

Fold Recognition

Our survey of known protein structures has allowed us to conclude that evolutionarily related proteins have similar structure. Does this observation mean that evolutionarily unrelated proteins always have different structures? The answer is no, because, very often, apparently unrelated proteins share a similar topology (Figure 47).

The explanation of this experimental observation can be that apparently unrelated proteins have diverged so much that our sequence-based methods are not sensitive enough to detect their true evolutionary relationship, or the specific topologies observed are energetically favored or easier to reach kinetically, and, therefore, more often used by nature. Whatever the reason, this observation implies that, even if our target protein does not seem to be

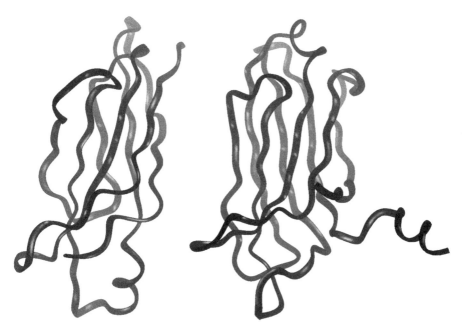

FIGURE 47
Two (probably) unrelated protein structures sharing a similar topology, a calcium/phospholipid binding protein (PDB id: 1RSY) from rat and the human growth hormone receptor (PDB id: 3HHR). These two proteins share less than 5% sequence identity, and yet their topology is extremely similar.

related to any known structure, a finite chance exists that it has a topology already present in our database of known protein structures. The most straightforward way to verify this possibility is to build several models for our protein by using as template every known structure, on the basis of every possible alignment, and then estimate whether one or more of our models seem "reasonable"; that is, whether they have a high sequence–structure fitness score that we need to define.

The search for an appropriate, and easy to compute, fitness function for this case is what prompted the development of the pairwise potentials discussed above. An extension of the dynamic programming algorithm, the double dynamic programming, can be used to obtain the optimal sequence–structure alignment. In the sequence alignment problem, we needed to optimize the sequence similarity score; here, we must optimize the overall pairwise potential. The algorithms must be more complex. In the sequence alignment problem, the score in each cell only depends on the amino acids in the two proteins and not on the path that we select at the end, whereas, here, the value of the function depends on the distance between all pairs of amino acids in the protein and, therefore, on which path we take. Indeed, double dynamic programming methods are quite slow and memory intensive. Thus, approximations are often used to simplify the task.

Another alternative method for recognizing a known fold that can fit a given protein sequence is based on the profile. We replace each amino acid of our target sequence with a symbol that indicates its propensity to be observed in a given structural environment. Usually, we take into account the propensity of the amino acid for being in one of the secondary structure types, for being in a hydrophobic or a hydrophilic environment, and for being more or less exposed to a polar solvent. This procedure recasts our sequence into a new sequence composed of a different set of characters. Whenever possible, we can use a multiple alignment of members of the same family of our target sequence because we know they share the same (although yet unknown) structure. Next, we can analyze each of the proteins of known structure. For every position, we do not take into account which amino acid happens to be present, but rather examine the property of the position (i.e., its secondary structure, its environment, and its exposition to the solvent) and assign a character that describes the observed combination of properties. Thus, our database of protein structures is represented by a set of strings.

The string that represents the query sequence, or its multiple-sequence alignment, can now be compared, by use of the same techniques described for the detection of evolutionary relationships, with each of the strings that represent the structures. As for sequence-based database searching methods, we need a background distribution to evaluate the significance of the scores that we obtain, and this distribution can be obtained by reshuffling our sequence or by creating reasonable "decoy" structures.

Fragment-Based Methods

What can we do if our protein does not seem to share any evolutionary relationship with a protein of known structure and if no significant score can be obtained by running the available fold recognition methods? We can try to predict specific features from the sequence and, for example, try to predict its secondary structure. However, recent developments in protein structure prediction might allow us to attempt the construction of a full-fledged, three-dimensional model of our protein, even in these difficult cases.

The "new-fold predicting methods" are usually fragment based; that is, they combine fragments of known structures, taken from our database of known protein structures, to construct a model of the target protein. Fragments with identical sequence can assume different conformations in different structures, so we cannot just search for fragments of known structure that have a sequence identical, or similar, to some fragment of the target protein and join them together. However, the innovative idea behind the new fold methods is that the distribution of conformations in which we find a fragment with a given sequence can be related to the propensity of that sequence to assume each of these conformations. We can retrieve all fragments sharing some local sequence similarity with each of the fragments of our target protein and join them in many combinations. This procedure generates a large but finite set of putative models that we can optimize by application of genetic algorithms, for example. The problem is reduced to a search for the "best" model among a given finite set of conformations, and we can use a sequence–structure fitness score to rank our models.

These methods have raised an enormous interest because they seem to be the only current way to obtain (although not always) a full, three-dimensional model of a protein that has no sequence or structure relationship with the set of known proteins. These methods can also be used to design new protein structures.

Natively Unfolded Proteins

The assumption that protein function requires structural organization does not necessarily mean that all proteins in a cell have a defined structure at all times. In X-ray crystallography, often no electron density can be observed for some regions of the analyzed protein, and these anomalies may correspond to disordered regions. Analogously, nuclear magnetic resonance experiments can indicate that regions of a protein fluctuate in the solvent rather than remaining fixed in a unique conformation. Spectroscopic techniques can highlight a protein that lacks a defined secondary structure or, although the secondary structure is present, that does not show the asymmetric interactions characteristic of a folded protein. The hydrodynamic dimension of a protein can be measured and compared with that of typical native globular proteins with corresponding molecular mass.

With the use of these techniques, a growing number of proteins have been found to share the intriguing property of being partially or completely unfolded. The proteins are called "natively unfolded," "intrinsically unstructured," or "intrinsically disordered."

The "disorder" property is conserved through evolution, which suggests that it has a functional significance. Intrinsic flexibility seems to be an important prerequisite for molecular recognition, and intrinsically disordered proteins are involved in a number of important biological processes, such as cell cycle control, transcriptional and translational regulation, modulation of activity, and assembly of other proteins. Most of these proteins undergo a disorder-to-order transition to perform their function, and this property is advantageous for several reasons.

An intrinsically flexible protein might bind several targets, or it can provide a larger interacting surface in big complexes. Such proteins might also have been optimized to bind their targets with low affinity (as they have to "spend" energy for folding before binding). Moreover, the lifetime of an unfolded protein in a cell is probably shorter, which can serve as a regulatory mechanism.

Most likely, all these effects, and possibly others, are responsible for the existence of natively unfolded protein, but in any case, detecting which proteins or regions of proteins are intrinsically disordered has quite important implications, not only for functional assignment but also because natively unfolded proteins or proteins containing unfolded regions do not crystallize. Therefore, detecting them in advance can prevent structural biologists from wasting time and effort.

The prediction of disorder is a binary classification problem. It is perfectly suitable for automatic learning methods but is also approachable by use of statistical classifications based on sequence composition. Several publicly available methods reach respectable accuracy when tested on data sets derived from the PDB protein structure database. These methods provide estimates that more than 40% of the proteins from high organisms contain disordered segments longer than 50 amino acids and that more than 15% of the proteins are completely disordered. These estimates are based on predictions, and we cannot be sure that they do not contain systematic errors because of the data set used for training. In any case, these intriguing proteins seem to represent more than a negligible fraction of the proteome. Thus, predicting their characteristics and experimentally studying their properties is important.

Promising Avenues

Clearly, efforts should be focused on the ability to correctly rank a set of models for the same protein. Estimating their quality *a priori* would improve comparative modeling, fold recognition, and fragment-based techniques.

How do we accomplish this objective? We certainly cannot use very detailed energy calculations. In the case of comparative modeling based on close evolutionary relationships, the calculations would fail because they are not sufficiently precise to evaluate the energetic contribution of small deviations from the real protein structure. In all other cases, the calculations would not work for the opposite reason: the models are too far away from the real structure, and this distance would mask their relative differences. Less detailed energy terms, such as pairwise potentials, are more likely candidates, but we know experimentally that they are also not very good for ranking different models. Often visual inspection of protein models allows an expert to discriminate, at least to some extent, between good and bad models. Therefore, the problem is to encode the pattern recognition of the expert into an automatic method. This task is not easy. One should take into account several factors, which include the packing of the interior of the protein, the absence of unusual structural features and, certainly, the correlation of the model with known experimental data on the analyzed protein. Once again we stress how valuable the ability to automatically check the consistency of our predictions with known and experimental information embedded in the large body of available biological literature would be.

The prediction community has established a worldwide experiment (Critical Assessment of Techniques for Protein Structure Prediction, CASP) to evaluate the effectiveness of methods for protein structure prediction and highlight the bottlenecks of present methods. Every 2 years, crystallographers and nuclear magnetic resonance spectroscopists who are about to solve a protein structure are asked to make the sequence of the protein available, together with a tentative date for the release of the final coordinates. Predictors produce and deposit models for these proteins before the structures are made available, and, finally, a panel of assessors compares the models with the structures as soon as they are available, evaluates the quality of the models, and draws some conclusions about the state of the art of the different methods. The results are discussed in a meeting of assessors and predictors, and the conclusions are made available to the whole scientific community via the World Wide Web and the publication of a special issue of the journal *Proteins: Structure, Function, and Genetics*. CASP experiments have been ongoing since 1994. Their results are based on a very large number of submitted models (thousands in each edition) and represent an invaluable data set for assessing the performance not only of structure prediction methods but also of methods for differentiating and ranking models of different quality.

Suggested Reading

Anfinsen, C.B., Harrington, W.F., Hvidt, A., and Lindstrom-Lang, K. Studies on the structural basis of ribonuclease activity, *Biochim. Biophys. Acta* 17, 141–142, 1955.

Levinthal, C. Molecular model-building by computer, *Sci. Am.* 214, 42, 1966.

Finkelstein, A.V. and Ptitsyn, O.B. *Protein Physics*, Academic Press, London, 2002.

Jones, T. A. and Thirup, S. Using known substructures in protein model building and crystallography, *EMBO J.* 5, 819–822, 1986.

Chothia, C. and Lesk, A. The relation between the divergence of sequence and structure in proteins, *EMBO J.* 5, 823–826, 1986.

Sali, A. and Blundell, T.L. Comparative protein modelling by satisfaction of spatial restraints, *J. Mol. Biol.* 234, 779-815, 1993.

Dunbrack, R., Jr. and Karplus, M. Backbone-dependent rotamer library for proteins: application to side-chain prediction, *J. Mol. Biol.* 230, 543 574, 1993.

Tramontano, A. (1995). The architecture of loops in proteins, in *Advances in Computational Biology*, Vol. 2, Villar, H.O., Ed., JAI Press, Greenwich, pp. 239–259.

Tramontano, A. and Morea, V. Assessment of homology-based predictions in CASP5, *Proteins* 53 (Suppl 6), 352–368, 2003.

Finkelstein, A.V., Badretdinov, A.Y., and Gutin, A.M. Why do protein architectures have Boltzmann-like statistics? *Proteins* 23, 142–150, 1995.

Finkelstein, A.V., Gutin, A.M., and Badretdinov, A.Y. Boltzmann-like statistics of protein architectures: origins and consequences, *Subcell. Biochem.* 24, 1–26, 1995.

Sippl, M.J. Knowledge-based potentials for proteins, *Curr. Opin. Struct. Biol.* 5, 229–235, 1995.

David, T. and Jones, J.M.T. Potential energy functions for threading, *Curr. Opin. Struct. Biol.* 6, 210–216, 1996.

Chothia, C. Proteins: one thousand families for the molecular biologist, *Nature* 357, 543–544, 1992.

Orengo, C. Classification of protein folds, *Curr. Opin. Struct. Biol.* 4(3), 429–440, 1994.

Bowie, J.U., Luthy, R., and Eisenberg, D. A method to identify protein sequences that fold into a known three-dimensional structure, *Science* 253, 164–170, 1991.

Jones, D.T., Taylor, W.R., and Thornton, J.M. A new approach to protein fold recognition, *Nature* 358, 86–89, 1992.

Bonneau, R., Tsai, J., Ruczinski, I., Chivian, D., Rohl, C., Strauss, C., and Baker, D. Rosetta in CASP4: progress in ab initio protein structure prediction, *Proteins Suppl.* 5, 119–126, 2001.

Dunker A.K., Brown, C.J., and Obradovic Z. Identification and functions of usefully disordered proteins, *Adv. Protein Chem.* 62, 25–49, 2002.

Moult, J., Pedersen, J., Judson, R., and Fidelis, K. A large-scale experiment to assess protein structure prediction methods, *Proteins* 23, ii–v, 1995.

Tramontano, A. Of men and machines, *Nat. Struct. Biol.* 10, 87–90, 2003.

Problem 5

Membrane Proteins

Introduction to the Problem

Membranes separate cells or cellular compartments from the environment. However, characterizing a membrane simply as a passive barrier is misleading. On the contrary, membranes are complex dynamic structures whose components are continuously synthesized and degraded. All biological membranes are made of lipids and protein molecules held together by non-covalent interactions. The relative abundance of the components varies, depending on function. The plasma membranes of animal cells are approximately 50% protein in weight. The inner mitochondrial membrane, which is involved in energy transduction, is about 75% protein, whereas myelin, which functions as an insulator, only contains roughly 25% protein.

Membrane proteins perform many extremely important functions. As transporters, they carry molecules and ions into and out of cells. As receptors, they recognize external stimuli. They have enzymatic activity and are responsible for cell-to-cell communication. The enormous interest in these proteins, especially for therapeutic purposes, is, therefore, not surprising.

Unfortunately, the structure and function of membrane proteins is extremely difficult to predict, at least as difficult as predicting the structure of globular proteins. This difficulty might seem surprising because the membrane environment limits the possible conformations a protein can assume. However, this potential advantage is more than counterbalanced by at least two properties. First, membrane proteins interact with the membrane hydrophobically, and these interactions are nonspecific and, therefore, difficult to model. Second, the experimental determination of the structure of membrane proteins is much more difficult than determination of the structure of soluble proteins, and, as a result, the available number of solved membrane protein structures is very low (less than 1% of the total number of solved protein structures, although 20% to 25% of all proteins are estimated to be membrane proteins), and this low number limits the possibility of extracting useful rules about their sequence–structure relationship.

The Structure of the Membrane

The peculiar properties of membrane proteins are dictated by the environment in which they exist. Therefore, we must discuss the general features of a membrane. The most abundant lipids in the membrane are phospholipids (Figure 48), which are made up of a hydrophilic head and a hydrophobic tail. In a hydrophilic environment, phospholipids spontaneously arrange into a double layer, with the heads pointing towards the solvent and the tails packing against each other (Figure 49). The energetic reason is always the same: exposing a hydrophobic molecule to a polar environment is entropically unfavorable.

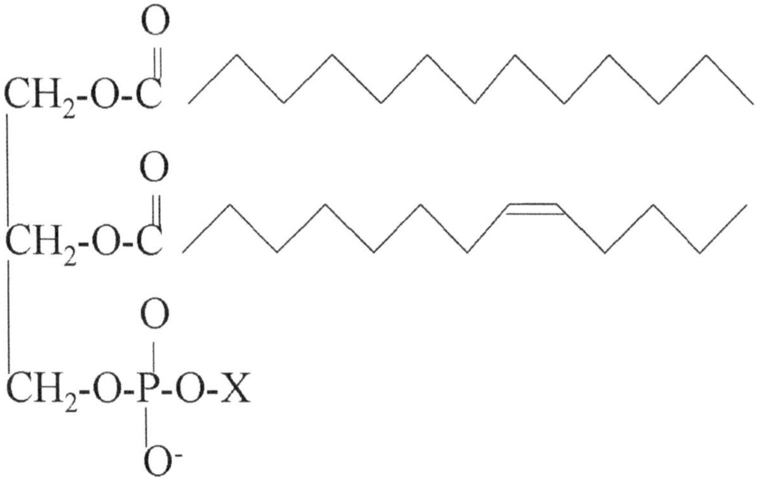

$$
\begin{array}{l}
\text{O} \\
\| \\
CH_2\text{-O-C} \\
| \\
\text{O} \\
\| \\
CH_2\text{-O-C} \\
| \\
\text{O} \\
| \\
CH_2\text{-O-P-O-X} \\
| \\
\text{O}^-
\end{array}
$$

FIGURE 48
The general structure of a phospholipid.

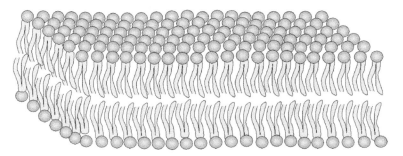

FIGURE 49
Schematic representation of a double layer of phospholipids.

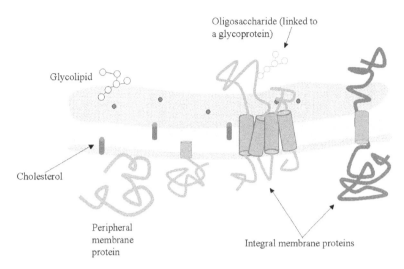

FIGURE 50
Schematic illustration of a membrane that contains peripheral and integral proteins.

The head group, designated by X in Figure 48, is different in different phospholipids and is attached to two fatty acid tails that can also differ in different membranes. However, in general, one tail does not contain double bonds and the other does. This difference is important because the rigidity of a double bond keeps the two tails from tightly packing against each other and, thus, maintains the membrane sufficiently fluid at body temperature. Individual lipid molecules are able to diffuse freely within the bilayer, and the embedded proteins are also able to move around. The membrane also contains another lipid, cholesterol, whose concentration may vary significantly. The cholesterol molecule inserts itself in the membrane with the same orientation as the phospholipid molecules. In this way, it immobilizes the first few hydrocarbon groups of the phospholipid molecules, which makes the lipid bilayer less deformable and permeable to small water-soluble molecules. Glycolipids are also found in membranes, with their sugar groups projecting into the extracellular space. Glycolipids can be protectors, insulators, and sites of receptor binding.

The Structure of Membrane Proteins

Some membrane proteins are partially inserted into the membrane, with a domain where hydrophobic residues are exposed on the surface in contact with the inner, hydrophobic part of the lipid bilayer and a polar domain interacting with the solvent. Others have a hydrophilic domain at each end of the protein and a single hydrophobic domain that spans the whole

FIGURE 51
Examples of integral membrane proteins: a channel protein, a porin, and the photoreaction center.

membrane. In the latter case, the polypeptide chain may traverse the membrane one or more times.

Another class of membrane proteins, the peripheral membrane proteins, is not inserted into the membrane and has no well-defined hydrophobic surface. These proteins are bound to the membrane principally by ionic associations with the polar phospholipid heads, by other membrane proteins, or by covalent links to glycolipids (Figure 50).

Three integral membrane proteins are shown in Figure 51 (see color insert after page 40): a bacterial potassium channel, the OmpF porin from *Escherichia coli*, and the photoreaction center I from a cyanobacterium. We will use these three as paradigms for highlighting the properties of this class of proteins.

The bacterial potassium channel has a rather simple architecture, and its role is to selectively allow ions to go through the membrane and, thereby, facilitate rapid signaling to every part of the cell. This mechanism forms the basis of muscle contraction, in which the signal is given by a flow of calcium ions, and of nerve signaling, in which the potassium channel is involved. A resting nerve cell has a high concentration of internal potassium, built up by selectively letting potassium in and calcium out, which creates an electrical potential difference. When the cell needs to signal other cells, the sodium channels open, and the influx of sodium depolarizes the cell. Later, the potassium channel re-establishes the potential difference. When a channel opens, the message is immediately felt by other channels of the same type, which also open. Thus, the signal is quickly transmitted.

The channel has to distinguish a potassium ion from a sodium ion. Therefore, the channel is formed by a selectivity filter that recognizes the two different ions and a gate that opens or closes according to the need. The filter is efficient: only one molecule of sodium escapes the selection mechanism for every ten-thousand potassium ions that pass through the channel.

Structurally, the channel consists of four chains that span the membrane and create an internal pore. Each chain is formed by α-helices that are roughly perpendicular to the plane of the membrane. The α-helices tilt at different angles and have different lengths.

The OmpF porin is a member of the β-barrel membrane protein class. These molecules are found in the outer membrane of gram-negative bacteria, where they either mediate nonspecific transport of ions and small molecules that passively diffuse though the pore or selectively allow the passage of molecules such as maltose and sucrose. The porin shown in Figure 51 transports nutrients and waste products across the outer *E. coli* membrane. Note that, as with the α-helices of the potassium channel, the strands are not exactly perpendicular to the membrane plane and have different lengths.

Higher organisms have internal organelles—the mitochondria in animals and the chloroplasts in plants—that are the energy factory of the cell. These organelles are surrounded by two membranes, the inner one of which is impermeable. The energy gained by the passage of electrons along a chain of electron acceptors, the last of which is oxygen, has the effect of accumulating protons on one side of the internal membrane, and the proton gradient "drives" a motor whose rotation is responsible for the formation of adenosine triphosphate (ATP), an energy-rich molecule that fuels many biochemical reactions.

β-barrel membrane proteins are also present in the external membrane of mitochondria. Evidence suggests that these organelles are derived from an early symbiotic event between an ATP-producing bacterium and a higher organism cell. In the mitochondria membrane of higher organisms, β-barrel membrane proteins appear to be involved in voltage-dependent channels.

The membrane-spanning region is formed by α-helices in the potassium channel synthetase and by β-strands in porins. Yet, the two structural organizations have a common feature that is directly related to a property of membranes. The surface of the membrane proteins is in contact with the apolar part of the lipid bilayer, and the side chains of hydrophobic amino acids can provide favorable interactions with this environment. However, the main chain contains polar atoms that cannot form hydrogen bonds with the surrounding hydrocarbon atoms and must bond with other atoms of the protein chain. (In the unfolded state, the polar backbone atoms form hydrogen bonds with the polar cellular environment, and folding is energetically unfavorable if they cannot form approximately the same number of hydrogen bonds in the folded state.) The backbone polar atoms of an α-helix form hydrogen bonds with other backbone atoms. Therefore, if a region of a protein traverses the membrane as an α-helix, all its polar atoms can satisfy their hydrogen bond potential. Another way of saturating potential hydrogen bond donors and acceptors of the backbone is to form a β-sheet, so that the last strand pairs with the first, forming a barrel, as in the case of the porin.

The photosystem I from a cyanobacterium is a multimeric protein that provides the machinery for converting the energy from sunlight into the chemical energy organisms need to perform mechanical, chemical, and osmotic work. It captures the photons and uses their energy to build sugar. The protein complex contains more than 100 cofactors; that is, small, organic molecules exposed around the edges or buried inside. Our photosystem I protein contains many small molecules, such as chlorophyll and carotenoids. Chlorophyll absorbs blue and red light, and this characteristic explains the green color that we see all around us in plants, whereas carotenoids are orange and mainly absorb in the blue region.

Prediction of the Structure of Membrane Proteins

Many methods of detecting transmembrane segments involve looking for stretches of hydrophobic residues that have a high tendency to form helices. However, membrane proteins can be made by helices or sheets and can form channels. Therefore, they can contain hydrophilic residues that can be accommodated on the surface of the protein that faces the channel. Furthermore, even if we detect helical transmembrane regions, we still face the problem of predicting their precise orientation and inclination with respect to the lipid bilayer.

We can separate the membrane protein structure prediction problem into three steps: (1) the prediction of their topography (i.e., the type and location of secondary structure elements), (2) the prediction of their topology (i.e., of the relative orientation of each of the elements with respect to the interior and exterior of the compartment enclosed by the membrane), and (3) prediction of their complete three-dimensional structure.

Prediction of the Topography of Membrane Proteins

The location of transmembrane helices can be predicted by searching the sequence of the target protein for stretches of 12 to 35 prevalently apolar amino acids that are connected by hydrophilic regions (the extracellular and intracellular regions) and that have a relatively high propensity to form helices. From a survey of the few known membrane protein structures and their homolog, we also know that the regions connecting the helices are usually shorter than 60 amino acids.

We must now define both the helical propensity and the hydrophobicity of each of the 20 amino acids. The helical propensity can be estimated by a statistical analysis of helix-forming residues in known protein structures, and various schemes give the amino acids numerical weights or rankings for their preferences. The simplest such scheme was devised by Chou and Fasman on the basis of the statistical distribution of amino acids in α-helices, β-sheets, and turns or loops in a set of known protein structures from the protein databank.

The quantitative measure of the propensity of an amino acid to be in an apolar environment is less straightforward. The hydrophobicity of each amino acid side chain can be experimentally determined by measurement (in cal/mol) of the change in free energy associated with the transfer of the molecule from water to an apolar solvent. However, the answer is not unique and depends upon the chemical form of the amino acid used in the experiments and upon the chosen apolar solvent. Do we use a peptide or a single amino acid? How do we take into account the contribution of the polar atoms of the main chain?

The first hydrophobicity scale, based on the relative solubility of amino acids in water and ethanol, was proposed by Nozaki and Tanford more than 30 years ago. They used isolated amino acids, which are not equivalent to residues in a peptide chain (the carbon and nitrogen main-chain atoms are free in an amino acid but not in a polypeptide chain). Furthermore, ethanol is not a very apolar solvent. Later, other investigators used octanol as solvent and blocked the amino acid termini with acetyl ($CH_3C=O$) and amide (NH_2) groups to mimic the situation of the residue within a protein or replaced the $C\alpha$ with a hydrogen atom, thereby using only the side chain. These approaches have a major drawback: if a polar atom of the side chain forms a hydrogen bond with an atom of the main chain, the hydrophobicity of both atoms increases, and this increase cannot be taken into account when modified amino acids are used.

Another way to estimate the hydrophobicity of an amino acid is to statistically analyze where it is found in protein structures. On average, hydrophilic amino acids are found more often on the surface of globular proteins, where they are accessible to solvent, than in the apolar interior of the proteins. To measure the extent of accessibility of an amino acid, we generally use the "accessible surface area," or ASA. ASA is the surface traced by the center of a water molecule as it rolls over the surface of the protein. The rationale behind using such a measure is that parts of the protein surface are not "buried" by other atoms, but are located in clefts of the protein that are too narrow to be contacted by the solvent (Figure 52).

A linear correlation exists between the extent of surface area that a residue or an atom exposes to solvent and its hydrophobicity. However, the proportionality constant is different for nonpolar, polar, and charged amino acids. We must, therefore, critically evaluate the likelihood of the hypotheses on which the ASA method is based. In this case, we assume that a protein interior is similar to a membrane environment. This assumption, however, is a rather crude approximation.

Attempts have been made to derive scales based on experimental and theoretical considerations about how well each amino acid would enter the lipid bilayer from an aqueous environment, and these scales work reasonably well for membrane proteins. The most commonly use scale is the one developed by Goldman, Engelman, and Steitz (called the GES scale).

Because of the difficulties in defining a reliable measure of hydrophobicity, many hydrophobicity scales can be found in the literature. These scales can

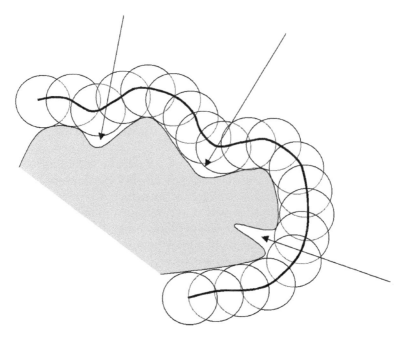

FIGURE 52

The solvent accessible area. Note the regions indicated by arrows. Although not buried by other, more external, parts of the structure, they cannot interact with a water molecule because they are located in nonaccessible clefts.

differ quite substantially (Table 5.1). Once we select a hydrophobic scale, we can identify membrane regions. Usually, we use a moving window (a concept described in the discussion of secondary structure prediction methods). The purpose is to average the hydrophobicity values of N adjacent residues. We must define the length of the window and a threshold value above which a stretch of residues will be predicted as transmembrane. Usually, N is around 20, and the threshold depends upon the selected hydrophobicity scale. Our confidence in the prediction can be increased by a high propensity of the region to be in a helical conformation (if we expect our protein to be helical) and by the observed length of the regions outside the membrane, which are usually shorter than 60 amino acids.

The detection of the likelihood that a stretch of residues is in a helical conformation can be based on the "moment analysis," defined as

$$\mu_H = \frac{1}{N} \left\{ \left[\sum_1^N H_n \sin(\delta n) \right]^2 + \left[\sum_1^N H_n \cos(\delta n) \right]^2 \right\}^{1/2} \tag{1}$$

TABLE 5.1

Some Commonly Used Hydrophobicity Scales

	Kyte and Doolittle	Hopp and Woods	GES
Alanine	1.8	−0.5	−1.6
Arginine	−4.5	3.0	12.3
Asparagine	−3.5	0.2	4.8
Aspartic acid	−3.5	3.0	9.2
Cysteine	2.5	−1.0	−2.0
Glutamine	−3.5	0.2	4.1
Glutamic acid	−3.5	3.0	8.2
Glycine	−0.4	0.0	−1.0
Histidine	−3.2	−0.5	3.0
Isoleucine	4.5	−1.8	−3.1
Leucine	3.8	−1.8	−2.8
Lysine	−3.9	3.0	8.8
Methionine	1.9	−1.3	−3.4
Phenylalanine	2.8	−2.5	−3.7
Proline	−1.6	0.0	0.2
Serine	−0.8	0.3	−0.6
Threonine	−0.7	−0.4	−1.2
Tryptophan	−0.9	−3.4	−1.9
Tyrosine	−1.3	−2.3	0.7
Valine	4.2	−1.5	−2.6

where H_n is the hydrophobicity of residue n, N is the number of residues in the segment, and the period can be either $100°$ (for α-helices) or $180°$ (for β-strands).

Because of the periodicity of the sine and cosine functions, a high hydrophobic moment is indicative of a periodicity of $360°/\delta$ in the value of H (i.e., of the hydrophobicity of the amino acids). A helix has a period of 3.6 and a strand of 2, which accounts for the values of δ. Therefore, if the amino acids on one face of the helix or on one side of the strand, respectively, are on average more hydrophobic than those on the other face, the computed moment will be high. Globular proteins contain several examples of amphiphilic helices, which are helices with one side facing the interior hydrophobic environment of the protein and one side facing the solvent. The hydrophobic moments of such helices are high, and this characteristic is sometimes useful in detecting their presence. The sequences of transmembrane helices are expected to have a high average hydrophobicity and helical propensity but a small moment because hydrophobic amino acids are expected to be present on every face of the helix. Unfortunately, the hydrophobicity pattern in bundles of transmembrane helices does not necessarily follow this rule. One side of a helix can face other helices of the bundle such that polar residues stabilize each other, which results in a higher hydrophobic moment.

The detection of secondary structure elements in β-barrel membrane proteins is even more complex. These proteins have a central pore that can accommodate polar residues. Some detection methods utilize hydro-phobicity plots that only count every other residue, often in conjunction with algorithms that predict the location of turns (i.e., of the regions between the strands) to define their boundaries.

In many cases, the prediction of the topography of a transmembrane protein can be simplified if the whole topology of the protein is considered, rather than just isolated secondary structure elements. For example, loops inside the compartment are more often positively charged than are loops outside (positive-inside rule), and this feature might aid in the location of helices. Other methods for predicting the topology of a membrane protein are discussed below.

Prediction of the Topology of Membrane Proteins

Although the extraction of rules from known examples of membrane pro-teins is made difficult by the paucity of available examples, automatic learn-ing methods have been applied to these cases and have yielded respectfully accurate results. The limitation of the data seems to be counterbalanced by the limited number of possible topologies.

Recent methods for predicting the topology of membrane proteins are based on Hidden Markov Models with five states (inside loop, inside helix tail, helix, outside helix tail, and outside loop) that are used to estimate the preference for each residue of the protein sequence to be in a transmembrane helix or in a loop. In some cases, dynamic-programming algorithms (similar to those we described for sequence alignment) are subsequently used to optimize the number and location of secondary structure elements. The positive-inside rule can be applied at the end of the procedure to select between the two possible orientations of the protein in the membrane.

Another approach to the prediction of membrane proteins is based on neural networks. Neural networks cannot be used directly for three-dimen-sional structure prediction. In the training phase, we must provide the neural network with both input and output data to derive the weights of the connections. For three-dimensional structure prediction, the input should be the sequence and the output should be the coordinates, but the coordinates of each of our training set examples depend on the specific frame of reference used. Similar three-dimensional motifs in two proteins have a completely unrelated set of coordinates, and, therefore, our network has no chance to generalize the rules relating one to the other. However, in the case of mem-brane proteins, we can utilize the plane of the membrane to predict the displacement of a residue along the axis of the barrel with respect to the membrane boundary. This procedure might help position the secondary structure elements in the correct orientation.

As in the case for secondary structure prediction, the exploitation of evolutionary information has been very beneficial for membrane topology prediction. Predicting the topology of a family of evolutionarily related membrane proteins is more effective than predicting the structure of a single sequence.

Prediction of the Three-Dimensional Structure of Membrane Proteins

No simple experiments provide even low-resolution data on the structure of membrane proteins, and this lack of data has constrained the possibility of devising methods for the prediction of the three-dimensional structure of these proteins. Here, we can only extrapolate from preliminary efforts.

After predicting the topography and topology of a protein, we can assemble the secondary structure elements by use of one of the arrangements of membrane proteins of known structure as template, but this approach is often misleading (like looking for the keys under a lamppost, not because we lost them there, but because it is the only place where there is enough light).

For β-barrel membrane proteins, prediction is even more complex. The large variety of functions that these proteins can perform is only matched by their very high structural variation. They can contain from 6 to 22 strands (apparently only an even number of strands is possible), and they can have different quaternary structure.

Techniques that we already described, such as fold recognition and molecular dynamics, have provided useful results in predicting membrane protein structures, more so than for globular proteins. We can impose more constraints for membrane proteins and, therefore, more effectively limit the search space. However, a word of caution is necessary: no accurate performance estimate is possible, given the low number of available structures. Therefore, we have very few clues about the appropriate methods to use and their reliability. This lack of data is probably the most serious obstacle to the improvement of methods.

Promising Avenues

Although a very limited number of examples of membrane protein structures are available, sequences are accumulating at a high rate. We must exploit this information even more than in the case of globular proteins. Surprisingly, however, very few attempts to develop homology detection methods specific for membrane proteins have been made. Very often, the substitution matrices derived from the alignment of globular proteins are used to search in databases for members of a membrane protein family, and they cannot represent a good model for the evolution of membrane proteins, because the structural constraints are different. The approach of combining several methods has been very beneficial in the search for globular protein structures. Preliminary attempts to use the same strategy for membrane proteins are encouraging.

Suggested Reading

Nozaki, Y. and Tanford, C. Solubility of amino acids and 2 glycine peptides in aqueous ethanol and dioxane solutions—establishment of a hydrophobicity scale, *J. Biol. Chem.* 246, 2211, 1971

Chou, P.Y. and Fasman, G.D. Prediction of the secondary structure of proteins from their amino acid sequence, *Adv. Enzymol. Mol. Biol.* 47, 45–148, 1978

Eisenberg, D., Weiss, R.M., and Terwilliger, T.C. The hydrophobic moment detects periodicity in protein hydrophobicity, *Proc. Natl. Acad. Sci. USA* 81, 140–144, 1984.

Rost, B., Casadio, R., Fariselli, P., and Sander, C. Transmembrane helices predicted at 95% accuracy, *Protein Sci.* 4, 521–533, 1995.

Chen, C.P. and Rost, B. State-of-the-art in membrane protein prediction. *Appl. Bioinformatics* 1, 21–35, 2002.

Rost, B. Prediction in 1D: secondary structure, membrane helices, and accessibility. *Methods Biochem. Anal.* 44, 559–587, 2003.

Mele, K, Krogh, A. and von Heijne, G. Reliability measures for membrane protein topology, *J. Mol. Biol.* 327, 735–744, 2003.

Problem 6

Functional Site Identification

Introduction to the Problem

In previous problems, we discussed some of the methods that can be used to identify protein residues that play functional roles. The evolutionary conservation of a sequence pattern is indicative of selective pressure, and, because natural selection acts on function, conserved residues are either directly responsible for function or essential for maintaining the structure that allows the correct positioning of functionally important residues. The analysis of conservation in protein sequences can be used to detect the functional sites of a protein. In this problem, we address the problem of how to detect functionally relevant residues when we know the three-dimensional structure of the protein.

Proteins perform a multiplicity of functions. Thus, when we say "functional residues," we refer to a variety of roles. In an enzyme, functional residues correspond to the active or to the recognition site. In other molecules, they form the binding surface for another macromolecule. In proteins involved in selective uptake of atoms or small molecules, they can be gating residues that open or close the entrance. In molecules involved in signal transduction, they modulate transmission of the signal. In this and the next problem, we describe methods to identify the role of these residues by analyzing a protein's three-dimensional structure.

The structural determination of a protein once represented the final step of its characterization, after it had been isolated on the basis of its function and extensively characterized. X-ray diffraction experiments were used to gain insight into the detailed mechanism of the catalyzed reaction or of the specific recognition pattern between the molecule under study and its cognate molecules. Now, structural genomics projects, aimed at determining the structure of as many proteins as possible, are producing structures of proteins whose functions remain to be discovered. We can be presented with cases in which the function of the protein is known and use the three-dimensional structure to deduce details about the role and interaction of specific atoms of the residues. We can also have cases of "structures without a history"; that is, we can be given a structure of a protein but no clue about the function. Such proteins are the ones we discuss here.

Structural Genomics

In Problem 4, we learned that an approximate model of a protein can be obtained on the basis of its evolutionary relationship with a protein of known structure. Every time we solve the three-dimensional structure of a protein, we provide information on the architecture of all the other proteins of the evolutionary family. The closely related proteins can be modeled with relatively good accuracy, but we also gain information about even the most distantly related proteins, albeit less detailed.

Thus, priority should be given determination of the structure of members of evolutionary families for which no structure is available, so that we can produce models for several other proteins, with the goal of compiling a menu of the possible protein architectures. Because we cannot solve the structure of every protein in the universe, this strategy is an effective use of resources. Worldwide projects that sample the structure space by determining the structure of proteins that belong to families for which no structural information is available are known as structural genomics projects. Structural genomics also include community-wide projects that focus on the structure determination of as many proteins as possible from the same organism, pathway, or biological process to use the information to derive a more complete view of the system.

The first step in a structural genomics project is the selection of the target proteins. Most of the computational biology techniques that we have described in the previous problems play a major role in this initial phase. The protein to be studied can be selected on the basis of its lack of similarity to proteins of known structure, so that it will provide us with the structural organization of a new evolutionary family, or it can be selected because it is predicted or known to be part of the target biological process.

The protein-coding gene first must be cloned in an appropriate vector, expressed, purified, and crystallized. The X-ray diffraction pattern of the crystal of the protein can then be used to solve its structure. The process should be as automated as possible. However, some of the steps are very difficult to standardize. The appropriate expression systems are different for different proteins, purification techniques are protein-specific, and the conditions in which different proteins crystallize are widely different. Therefore, not all the desirable protein structures can be determined with such an approach. Some still require focused and nonautomated efforts that are time consuming and labor intensive.

In projects designed to cover the structure space, the experimental complications are compounded by the bioinformatics difficulties of accurately detecting which proteins are unrelated to proteins of known structure. These projects rely on sequence identity measures in which any protein that shares less than a given percentage of identical amino acids with a protein of known structure, usually around 30%, is considered a good candidate and sent through the cloning, purification, and crystallization procedures. This process is too crude an approximation.

Pairwise sequence similarity can be used to deduce the expected similarity between the structure of the two proteins, but it is not the most effective way to detect evolutionary relationships. The transitive properties of evolutionary relationships allow us to use much more sensitive methods based on multiple-sequence alignments. Furthermore, we must estimate how appropriate the protein structure is as a template for constructing models of the other proteins of its family; that is, the expected reliability of the models we can build by using it as a template.

The quality of a comparative model depends upon the extent of structural divergence between the target and the template and upon the quality of the sequence alignment between the two protein sequences. The latter is usually derived from a multiple-sequence alignment of as many proteins of the family as possible, and its accuracy depends upon the number and similarity distribution of the sequences of the protein family. Furthermore, the quality of an alignment is not constant throughout the whole sequence, and different regions have different reliability. Both of these aspects are under extensive study at present and will certainly result in target selection strategies that will be more effective in sampling the "fold space" available to proteins.

Even if experimental difficulties prevent the structural determination of all our target proteins, many of them will appear in the database, often without being attributed a function and requiring analysis by computational techniques. Because structure is better conserved than sequence in evolution, we can hope that the comparison of the protein structure with that of all the other proteins of known structure might reveal evolutionary relationships that allow us to deduce function, much in the same way as we described in Problem 3.

Even if our protein is not evolutionarily related to any other protein for which we have functional information, its functional site might locally resemble a functional site of a protein of known function as a result of convergent evolution. Convergent evolution, the independent evolution of the same property in different species, is often observed both in organs and in specific protein structural features. Examples of the former are the development of a long, sticky tongue, few teeth, a rugged stomach, and large salivary glands in anteaters all over the world. These features evolved independently, like the wings of bats and birds and the eyes of invertebrates and vertebrates.

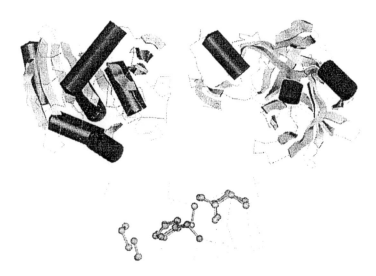

FIGURE 53
Comparison of the active sites of a protease and a lipase. The topology of the two proteins is very different (top), but the residues in their active site can be almost perfectly superimposed.

Arctic and Antarctic fish provide an example of convergent evolution at the protein level. These fish have developed a glycoprotein that circulates in the blood and lowers the temperature at which it freezes. The proteins evolved after the two groups separated.

Proteases are hydrolyzing enzymes that cleave a peptide bond. A class of these proteins uses a triad of amino acids: a serine, an aspartic acid, and a histidine. A completely different class of enzymes, the lipases, cleaves lipids, but has developed a strikingly similar active site, formed by the same three amino acids in the same relative position and playing the same chemical role as in serine protease (Figure 53) (see color insert after page 40), although they appear in two completely different protein architectures. This finding implies that, even in the absence of a global sequence or structural similarity between two proteins, local similarities can allow function to be detected.

Given the structure of a protein of unknown function, our first task is to compare it with all known protein structures to see whether any evolutionary relationship can be detected and, in this case, whether residues that correspond to known functional residues are conserved between the homologous proteins. Should this strategy fail, we can search for local similarities of functional patterns. In either case, we must solve the problem of measuring structural similarity, both local and global, between two protein structures and of searching the structure database for proteins that share a significant structural similarity with the target protein.

The database of known structures contains several redundant similar structures. We should first calculate an all-against-all structural alignment of the known structures and cluster them so that the query protein can be compared against representative structures of each cluster.

Structural Superposition

Root Mean Square Deviation

We previously introduced the definition of root mean square deviation:

$$rmsd = \sqrt{\frac{1}{N}\sum_{i=1}^{N}\left[(x_i - x_i')^2 + (y_i - y_i')^2 + (z_i - z_i')^2\right]}$$

where (x_i, y_i, z_i) and (x_i', y_i', z_i') are the coordinates of the atoms that we want to superimpose to each other.

If the correspondence between the pairs of atoms we want to superimpose is known, we can easily measure their rmsd and also calculate how to optimally superimpose them. We apply the rigid-body translation $T = (T_x, T_y, T_z)$ and rotation $R = (R_x, R_y, R_z)$ to one of the proteins that minimize the rmsd between the given set of atom pairs:

$$rmsd(T, R) = \min_{T,R}\sqrt{\frac{1}{N}\sum_{i=1}^{N}\left[(x_i - R_x x_i' + T_x)^2 + (y_i - R_y y_i' + T_y)^2 + (z_i - R_z z_i' + T_z)^2\right]}$$

Usually, however, we do not know which are the corresponding pairs of atoms. When we compare two protein structures with identical sequences, such as a structure and a model or two different structural determinations of the same protein in different conditions, we can superimpose each pair of equivalent atoms, but the result can be misleading. If one region of the model is incorrectly positioned with respect to the rest of the structure, the rmsd can be very high and a superposition might fail to highlight the similarity of the other regions of the structure. Exclusion of some regions of the protein that are locally different might be warranted to obtain a more informative superposition. In other words, we must decide whether we prefer to have more equivalent points at the expense of a lower rmsd and how much "quality" we are prepared to lose to achieve a superimposition that includes more atoms of our proteins. The optimal superposition of two protein structures can only be defined if we decide either the minimum number of atoms that we want to superimpose on each other or the maximum value of rmsd that we are prepared to accept.

Figure 54 shows a hypothetical plot for the superposition of two different models of the same protein with the experimental structure. The rmsd values are plotted as a function of the number of atoms included in the superposition, and they vary according to the superimposed fraction. The model that corresponds to the thick line has a lower rmsd deviation from the structure than does the model that corresponds to the thinner line when about three fourths of the structure is considered. The situation is reverted

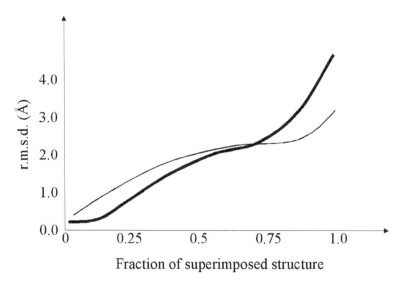

Fraction of superimposed structure

FIGURE 54

Comparison of the structure of two hypothetical models for the same protein. The percentage of superimposed Cα is shown on the x-axis, and the resulting rmsd is shown on the y-axis. Notice that one model (thick line) is closer to the experimental structure for 75% of the structure, whereas the other (thin line) provides an overall better model if the whole structure is taken into account.

if the entire structure is considered. The second model has an overall lower rmsd, and indeed the two lines cross each other.

This observation illustrates the problem of unambiguously defined structural similarity, even between three-dimensional structures with the same sequence (and also highlights the problems that result from assessing models in the worldwide CASP experiment). The issue becomes even more relevant when we address the problem of measuring similarities between two different protein structures.

Although rmsd is a very commonly used measure of distance, it is not the only one and, in some cases, might not be the most suitable to highlight common regions of two proteins. One alternative is to count the number of atoms within a certain distance threshold after superposition of the two structures. This strategy is often used to compare models with their respective experimental structures. If a region of the protein is grossly incorrect in two different models, the rmsd value will take into account how distant from the experimental structure each model is, although this information is not necessarily useful from a biological point of view.

Structural Superposition between Two Different Proteins

To measure the structural distance between two different proteins, we must decide the pairwise correspondence between their atoms. Paired atoms

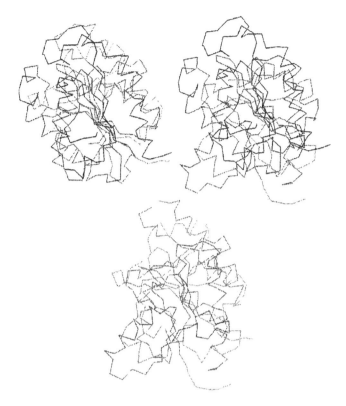

FIGURE 55
Different superpositions of evolutionarily related proteins: both superpositions have an rmsd of about 3 Å, but the one on the right includes 64 residues and the other includes only 36.

should be chemically equivalent (e.g., Cα carbons of one protein should correspond to Cα carbons of the other protein). Therefore, when we super-impose the structures of proteins with different amino acid sequences, we can only refer to Cα, to main-chain atoms, or, in some cases, to main-chain atoms plus the first carbon atoms of the side chain (Cβ) in nonglycine residues. The last possibility is useful because it takes into account the direction of the side chains.

As for sequence alignment, ideally the superposition should be such that corresponding residues (i.e., pair of residues close to each other in the two structures) are those that have evolved from the same residue of an ancestral protein. We must make sensible choices that will increase the likelihood that our structural alignment is biologically significant.

Given the complexity of the problem, ambiguities are expected to arise. Figure 55 (see color insert after page 40) shows different structural superpo-sitions of C atoms between two evolutionarily related proteins, a cutinase (PDB id:1CUT) and a low-molecular-weight phosphotyrosine protein phos-phatase (PDB id:1PNT). Cutinase catalyzes the hydrolysis of cutin, a polyester that forms the structure of plant cuticle and allows pathogenic fungi to penetrate into the host plant during the initial stage of the fungal infection, whereas the phosphatase might play a role in the function of synapsis. Even when the rmsd values are the same, the optimal superposition can be different.

Some methods that compute the corresponding superposition between two protein structures start from an initial list of pairs of "seed residues" and iteratively refine the list to maximize the number of pairs and minimize their distance. The seed residues can be superimposed by calculating the translation and rotation matrix needed to minimize their rmsd. The resulting structural superposition can be used to derive a larger set of equivalent residues. We associate each residue of one protein with the closest residue in the other protein after the initial superposition, repeat the superposition step, and obtain a new list of equivalent residues. The procedure can be iterated until the total rmsd does not decrease and the number of matched residues in the two proteins does not increase. The results of this type of procedures are dependent upon the choice of the seed residues. Therefore, typical approaches start from more than one initial set, optimize the corresponding final alignment, and choose the best one at the end of the procedure.

If the two proteins share some similarity in sequence, the sequence alignment of conserved regions can be used as a starting approximation of the pairing (we need at least three pairs of residues to start). Alternatively, one can choose pairs of short fragments that have an easily detectable structural similarity, or, if none of the above applies, one can start from a set of randomly selected pairs of residues.

The genetic algorithms we described in Problem 4 are also suitable for solving the optimization problem of structural superposition. They construct an initial population composed of a large number of random superpositions between the two proteins (i.e., a large set of randomly assigned residue pairs) and evolve it via mutations and crossing-over by use of the rmsd as fitness function.

Distance Matrices

Other very commonly used techniques for structural superposition utilize the distance or contact matrix, a concept that is also useful in several other applications. In a distance matrix, each column and each row of the matrix represents an amino acid of a protein, and each cell contains the intermolecular distance between the amino acid of the row and the amino acid of the column. The distance can be calculated between $C\alpha s$ or $C\beta s$, depending on the specific application. Selection of a cutoff distance value allows the matrix to be visualized as in Figure 56 (see color insert after page 40), in which every cell that contains a distance lower than the cutoff is filled. The contact matrix for an SH3 domain (PDB id:1RUN) is shown in the figure using a cutoff of 8 Å. Amino acids close in the sequence are in contact and, hence, the diagonal of the matrix is surrounded by filled cells. The geometry of α-helices places amino acids three or four positions apart in contact. Therefore,

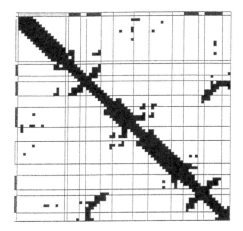

FIGURE 56
The distance matrix for an SH3 domain (extracted from the PDB entry 1RUN), the structure of which is shown on the right. Cells containing distances lower than 8 Å are filled. The secondary structure of the protein is shown as gray (helices) and black (strands) bars in the first column and in the first row.

regions of α-helices show up as diagonal lines. A pair of parallel strands has contacts between consecutive residues and also appears in the matrix as regions parallel to the main diagonal. In antiparallel pairs of strands, residues are in contact in reverse order (e.g., residues 4, 5, 6, 7, and 8 are in contact with residues 18, 17, 16, 15, and 14, respectively) and are represented by diagonals orthogonal to the main diagonal. Long-range contacts between different parts of the molecule show as off-diagonal patches of filled cells.

Proteins with similar three-dimensional structures have similar sets of interresidue distances and, hence, similar distance matrices. If they share a local structural similarity, this feature is reflected in the presence of a similar pattern in the distance matrix, and the detection of similar submatrices in the two matrices can provide the starting point for the structural superposition. This strategy is used in the very popular DALI method.

The first step of the algorithm splits the distance matrices into overlapping submatrices of fixed size (six consecutive residues on each protein) and searches for matching contact patterns within a given threshold; that is, for regions of the first protein that are similar to regions of the second protein. The second step consists in finding pairs of submatrices in the first protein that are similar to pairs of submatrices in the second protein. Assume that the submatrices $M_{1,j}$ and $M_{1,k}$ of the first protein match the submatrices $M_{2,u}$ and $M_{2,v}$ of the second protein, respectively. DALI calculates the rmsd between the fragments corresponding to $M_{1,j}$ and $M_{1,k}$ in the first protein and those corresponding to $M_{2,u}$ and $M_{2,v}$ in the second. If the rmsd is within an allowed threshold, the superposition is accepted. From this list of "pairs of pairs," the algorithm can construct a chain of connected contact patterns by joining pairs that share a submatrix. If the regions that correspond to the pairs $M_{1,k}$ and $M_{1,l}$ of the first protein can be superimposed on the regions that correspond to $M_{2,v}$ and $M_{2,z}$ in the second protein, then the regions that correspond to $M_{1,j}$, $M_{1,k}$, and $M_{1,l}$ in the first protein will be superimposed on those that correspond to $M_{2,u}$, $M_{2,v}$, and $M_{2,z}$ in the second.

DALI uses a Monte Carlo simulation to search for the best combination of matches among mutually exclusive sets of matching regions. In practice, DALI also uses clustering techniques to reduce the number of pairs of sub-matrices. This method is the basis for the database of structurally related proteins named FSSP (Families of Structurally Similar Proteins).

CATH is another database of structurally similar proteins, based on a different algorithm for structural comparison called SSAP. The general idea in this case is to generate a matrix that contains the vectorial distance between every pair of amino acids. For each pair of rows of the two matrices (i.e., for each pair of residues), SSAP computes a difference matrix. The difference matrix for row h of the first protein and row k of the second contains, in the cell i,j, the difference between the vectors that connect residues h and i of the first protein and residues k and j of the second, converted to a scalar similarity value. The best path through each of these matrices is computed by a dynamic-programming algorithm similar to that used for sequence alignment to construct a "summed scoring matrix." The best path in this final matrix is used to determine the structural alignment.

Other structural superposition methods are based on matching secondary structure elements between two proteins to find the best superposition between these elements and on subsequent refinement of the superposition (VAST, PrISM) or are based on the identification of locally similar fragments (MAMMOTH).

Some methods are based on graphic representation of protein structures in which each residue corresponds to a node, and the edges represent contacts between different residues that are within a predefined threshold. The superposition is obtained by finding subgraphs (i.e., subsets of the vertices and edges) of the original graphs that represent the two proteins that are isomorphic to each other (i.e., have the same number of nodes connected in the same way). These types of algorithms are usually slow and can only be used for pairwise comparisons.

The methods we described are useful for detecting a global structural similarity between two proteins, and they can provide a list of equivalent residues between two apparently evolutionarily unrelated proteins. The analysis of the conservation of these equivalent residues, as well as the presence of rarely observed structural features shared by the two proteins, can highlight an evolutionary relationship, and the extent of conservation of residues known to be involved in function can help determine the likelihood that the two proteins share the same function.

The complexity of the superposition problem leads to the development of algorithms that are very time consuming. Therefore, the community has precompiled classifications of similar protein structures that are extremely useful and widely consulted.

Structural Classification of Proteins

Along with the two databases for structural classification of proteins we have already mentioned, FSSP (based on the DALI algorithm) and CATH (based on the SSAP method), is a third widely used database called SCOP. All these structural classification systems are domain based because different domains of the same protein can belong to different structural classes.

SCOP has a hierarchical organization that includes class, fold, superfamily, and family. The main class types in SCOP are all α, all β, α plus β, and α/β. The distinction between the last two classes depends upon the relative arrangement of the α-helices and the β-strands. In the α plus β class, the two regions of secondary structure are somewhat separated in the protein, whereas in the α/β class, they pack against each other. A protein is assigned to one of the classes according to the predominant type of its secondary structure. The other classes include multidomain, membrane and cell surface proteins and peptides, small proteins, peptides, designed proteins, and low-resolution structures. The fold classification includes proteins with similar topological arrangements in which an evolutionary relationship cannot be identified. The superfamily includes proteins that are believed to share a common ancestor. If the sequence similarity between two or more proteins clearly points to an unambiguous evolutionary relationship, they are grouped in the same family. The classification in SCOP is essentially manual, although some automatic preprocessing is employed to cluster clearly similar proteins.

CATH is also hierarchical but, as we mentioned, is based on an automatic classification that utilizes the SSAP method. Classification levels in CATH are class, architecture, topology, and homology. The class is defined, as in SCOP, on the basis of the predominant type of secondary structure: all α, all β, and $\alpha\beta$, and domains with little or no secondary structure. SSAP is used to define proteins that belong to the same topology on the basis of their structural distance, and the homology grouping is based on sequence comparison. The architecture classification is assigned on the basis of visual inspection of the proteins and on the basis of literature data.

As we mentioned, FSSP is completely automatic and does not directly assign hierarchical levels to protein structures. A pairwise comparison of the structures of proteins sharing less than 25% sequence identity with any other in the PDB database is computed. Next, for each protein, the average and standard deviation of its similarity value distribution with all other proteins is calculated. The final score for a similarity is given as a Z-score (this score is the difference between the observed value and the average value in the total distribution divided by the standard deviation). Protein pairs that have a Z-score higher than 2.0 are deemed to be structurally similar, and they are arranged in a tree that uses the Z-score as a measure of distance.

None of these methods appear to be intrinsically better than the others, and they mostly agree with each other, although a precise comparison is not straightforward because of, among other things, the uncertainty in the precise definition of domains. The manual assignment in SCOP has the drawback that its updates are less frequent than one might wish. On the other hand, automatic methods all have limitations and fail in some cases, especially in difficult borderline cases. Therefore, the user has to exercise some judgment and, as always, assess which method is most appropriate for the specific task.

Detecting the Active Site

Perhaps the most general way to define functional regions in proteins is that they bind other molecules, they are the substrates in active sites of enzymes, and they are macromolecules in antibodies and regulatory proteins. Can we detect active sites on the basis of the knowledge of a three-dimensional structure? Enzyme substrates are relatively small molecules, and the protein is expected to present a cleft for binding them. On the basis of a few experimental observations, we know that in multidomain or multichain enzymes, the active site is often located between the domains or the chains, whereas in single-chain enzymes, the largest cleft is often the one that binds the ligand, more than 70% of the time. These observations are difficult to validate because of the difficulty in obtaining statistically significant and unbiased data sets.

In any case, we want to accurately and quantitatively describe a protein surface, to find "interesting" shapes or shapes that we have already observed in other enzymes. We also want to characterize physicochemical properties, such as electrostatic potential and hydrophobicity, of the protein surface.

The solvent-accessible surface described in Problem 5 is the most popular and effective method of describing a molecular surface. The majority of surface description and comparison techniques either use it directly or are based on some clever approximation of it that selects "significant" points of the surface that still retain the relevant shape information.

The calculated surfaces of proteins have been studied by an impressive variety of methods to detect depressions and knobs. Scientists have looked at the area-to-volume ratio and at the possibility of fitting the shape of the protein to dodecahedrons or spheres, and they have used Fourier analysis, spherical harmonics, and contour maps.

Some methods of identifying local surface similarities have been based on vectors normal to the surface that are subsequently clustered to derive a set of significant vectors that can be compared between two proteins. Other methods embed the protein in a three-dimensional grid with each cube of the grid marked as internal or external to the protein and often labeled with a property such as hydrophobicity or charge. Subgraph similarity detection and geometric hashing techniques have also been used. In all these methods, we have a combinatorial explosion. The active site is usually formed by a few amino acids, generally fewer than five, and, therefore, the number of different residue combinations in the protein structure database that we need to compare with each other is enormous, and remains so even if we limit ourselves to only those combinations of residues "likely" to be part of active sites (i.e., that contain putatively reacting groups).

A different approach is to accurately survey known active sites and prepare a catalog of their features. These features can be the shape of the surface or the relative location and type of atoms involved in the enzymatic mechanism. The data are stored in databases and can be used to search for matches in a newly determined protein structure. This approach is probably the most effective, although it does require a large amount of manual inspection, and it can only recognize already known arrangements of functional chemical groups.

Despite all the efforts, the problem of identifying where action takes place in a protein structure is not yet solved and the reason, as so often happens in bioinformatics, should not be attributed to inadequacy of the algorithms themselves, but to the complexity of biological molecules. Protein surfaces are formed by exposed side chains that are intrinsically flexible, so any static description of a surface will fail to detect the details, which is where the action occurs.

Another relevant problem impairs our ability to detect function on the basis of the shape of the surface of the active site and is also relevant to predicting macromolecular interactions. Trypsin, a digestive enzyme of the serine protease family, can cleave a peptide bond after both lysine and arginine. Their side chains are recognized by the same pocket formed by the protein structure. How does the protein accommodate two different side chains in the same pocket? When lysine is bound, the enzyme recruits a water molecule that fills the space of the pocket that would otherwise be left empty, and, therefore, lead to a less energetically favorable interaction. Interfaces between molecules often contain buried water molecules, which challenges the seemingly reasonable hypothesis that a similar function requires a similar shape.

The problems affecting our ability to detect functional sites are, in practice, caused by the extreme versatility of protein structures. Their ability to finely modulate their activity is the result of their flexibility and ability to take advantage of the environment, which is what gets in the way of many automatic methods.

Moonlight Proteins

Evolution uses whatever is available. Therefore, the fact that several proteins are endowed with more than one function should not come as a surprise. A protein has a large solvent-exposed surface that is not under a strong selective pressure and can, thus, evolve and form pockets or sites that can be used for additional functions. These proteins are called moonlight proteins, and undoubtedly many more will be discovered also among those to which a function has already been attributed.

Both phosphoglucose isomerase and phosphoglycerate kinase, two enzymes involved in the metabolic conversion of glucose, have additional functions. Phosphoglucose isomerase binds to cell surface receptors and increases tumor cell mobility. Phosphoglycerate kinase reduces the disulfide bridges in a protein called plasmin and, in doing so, activates a cascade that leads to the production of angiostatin, a protein that inhibits angiogenesis (i.e., formation of new blood vessels). Neurophilin is a receptor present in both endothelial cells and in neurons. In endothelial cells, it senses the need for new blood cells. In neurons, it helps direct the axons in the right direction. The enzyme prostaglandin H_2 synthase has two active sites and catalyses two consecutive reactions in the conversion of arachidonic acid to prostaglandin, a mediator of inflammation. In birds, the same protein acts as crystalline in the eye and as lactate dehydrogenase.

The combination of functions found in moonlight proteins is very diverse and often part of different cellular processes. The factors that determine which of the functions the protein exhibits are also highly variable. They can include cellular localization, presence of a ligand and quaternary structure, or combinations of individual factors.

Moonlight proteins can provide advantages to the cell, such as coordinating related cellular activities. However, they are an additional headache for protein bioinformaticians, as well as for medical professionals, because correcting or inhibiting a target function of a moonlight protein can have undesirable and unexpected effects on other unrelated functions.

Promising Avenues

Structural genomics projects have changed our way of looking at the problem of function assignment. Previously, we concentrated on the possibility of detecting function from the amino acid sequence and the evolutionary relationships of a given protein or of all the proteins of a genome, in the unverified and unexpressed hypothesis that the difficulty laid in doing so without the availability of a detailed structure characterization. Structural genomics projects challenged this view and have indeed proved that the task of detecting function is difficult, even when the structure is known.

A combination of heuristic and chemical knowledge is required. The number of known examples will increase and provide us with more solved instances of the problem, but at the same time, we must develop smarter algorithms to make the computational problem tractable. Searching all the proteins of known structure for similar clusters of amino acids to detect common subsites potentially involved in function is computationally prohibitive. Therefore, we must use our chemical knowledge to reduce the number of patterns to be compared, and we can do so by only considering residues that are likely to be involved in chemical reactions because of their chemicophysical properties. Alternatively, we can use evolutionary information to discard nonconserved residues. Some methods are indeed following this route, but how effective they will be is still unclear. As usual, the problem lies in the functional flexibility of proteins. Enzymes from the same superfamily do not necessarily preserve specificity. Most often, they do not: only 25% of the known superfamilies share a common function. Even the reaction chemistry in evolutionarily related enzymes is not necessarily conserved, and, even when it is, different residue types can play the same role in different enzymes.

This problem will occupy protein bioinformaticians for some time. This forecast is supported by the fact that worldwide initiatives for testing blind predictions of function are being initiated. In particular, the very popular CASP experiment, initially only devoted to structure prediction, is now challenging the participants to also attempt function prediction of the available targets. The results will be particularly instructive, as the experimental structure of the target proteins will be available and we will be in the position of estimating to what extent their knowledge will change, improve, or disprove the predictions.

Suggested Reading

Pastore, A. and Lesk, A. Comparison of the structures of globins and phycocyanins: evidence for evolutionary relationship, *Proteins* 8, 133–155, 1990.

Gibrat, J-F., Madej, T., and Bryant, S.H. Surprising similarities in structure comparison, *Curr. Opin. Structural Biol.* 6, 377–385, 1996.

Holm, L. and Sander, C. Dali: a network tool for protein structure comparison, *Trends Biochem. Sci.* 20, 478–480, 1995.

Holm, L. and Sander, C. The FSSP database of structurally aligned protein fold families, *Nucleic Acids Res.* 22, 3600–3609, 1994.

Murzin, A., Brenner, S., Hubbard, T., and Chothia, C. SCOP: a structural classification of proteins database for the investigation of sequences and structures, *J. Mol. Biol.* 247, 536–540, 1995.

Orengo, C.A., Michie, A.D., Jones, S., Jones, D.T., Swindells, M.B., and Thornton, J.M. CATH: a hierarchic classification of protein domain structures, *Structure* 5, 1093–1108, 1997.

Zemla, A. LGA: A method for finding 3D similarities in protein structures, *Nucleic Acids Res.* 31, 3370–3374, 2003.

Ortiz A.R., Strauss C.E., and Olmea O. MAMMOTH (matching molecular models obtained from theory): an automated method for model comparison, *Protein Sci.* 11, 2606–2621, 2002.

Taylor, W. and Orengo, C. Protein structure alignment, *J. Mol. Biol.* 208, 1–22, 1989.

Hubbard, T.J. RMS/coverage graphs: a qualitative method for comparing three-dimensional protein structure predictions, *Proteins Suppl.* 3, 15–21, 1999.

Godzik, A. The structural alignment between two proteins: Is there a unique answer? *Protein Sci.* 5, 1325–1338, 1996.

Brenner, S.E. and Levitt, M. Expectations from structural genomics, *Protein Sci.* 9, 197–200, 2000.

Blundell, T. and Mizuguchi, K. Structural genomics: an overview, *Prog. Biophys. Mol. Biol.* 73, 289–295, 2000.

Jones, S., and Thornton, J.M. Searching for functional sites in protein structures, *Curr. Opin. Chem. Biol.* 8, 3–7, 2004.

George, R.A., Spriggs, R.V., Thornton, J.M., Al-Lazikani, B., and Swindells, M.B. SCOPEC: a database of protein catalytic domains, *Bioinformatics* 20 (Suppl 1), I130–I136, 2004.

Problem 7

Protein–Protein Interaction

Introduction to the Problem

Most biological functions are mediated by protein interactions. These interactions can be physical, such as when two proteins form a complex, or "logical," such as when one or more proteins control the behavior of one or more other proteins without physical interaction. Examples of physical interactions are stable complexes, in which the functional unit is formed by more than one protein chain, as in the case of the glycogen phosphorylase enzyme, and transient associations, in which the protein chains are stable by themselves but can also interact to transmit a signal or as a response to external conditions. In logical interactions, one protein affects another protein by, for example, regulating its expression or changing the concentration of a factor that, in turn, is sensed by the target protein. The two modes of interaction are not exclusive. The same proteins can interact both physically and logically.

Detecting which proteins interact, how they do so, and what function is performed by their complex is at least as important as predicting the three-dimensional structure of individual proteins. Historically, the problem of detecting the site where two proteins (or a protein and a nucleic acid molecule) interact has been treated separately from the problem of finding where a small molecule such as an inhibitor or a substrate binds to a macromolecule. The two cases are similar only at a first glance; they differ quite substantially in the details of the techniques. Here, we discuss the macromolecular interaction problem. In the next problem, we briefly describe some aspects of the problem of docking small molecules to a macromolecule.

Protein Interactions

Metabolic pathways provide us with many examples of logical interactions. The concentration of a product is often "sensed" by other proteins in its synthetic cascade and modulates their activity. The presence of hormones is detected by cell surface receptors and transmitted to other proteins in the cell that can interact with the genetic material to activate or repress genes. External stimuli, such as the presence of food or poison, is sensed by a bacterium and transmitted to its flagella to direct the cell towards or away from the region of highest concentration of the sensed substance. These logical interactions can coexist with physical interactions. For example, hemoglobin senses the binding of oxygen and transmits the information from one of its subunits to the others via physical interaction. Other examples can be found in cell surface receptors. These molecules have an extracellular domain, a membrane domain, and an intracellular domain. Binding of a ligand to the extracellular domain can cause these molecules to form dimers (i.e., to associate with another receptor chain). The association of the corresponding intracellular domains allows the molecule to transmit the signal inside the cell (Figure 57).

Physical interactions can be stable or transient. The association of the two chains in alcohol dehydrogenase is stable; the chains act together to perform the protein's function. Hemoglobin also forms a stable complex made up of four chains (two identical α-chains and two identical β-chains). Hemoglobin is an α-helical protein and contains four heme groups, one for each chain (Figure 58) (see color insert after page 40). Each heme contains one iron atom at the center. The task of this protein is to transport oxygen from the lungs to the tissues; that is, from a region where the concentration of oxygen is high to a region where the concentration is low. None of the amino acid side chains is able to bind oxygen reversibly. Therefore, the protein uses an iron atom (Figure 59). The iron can form six chemical bonds: four with heme nitrogen atoms, one with the side chain of a histidine of the protein, and one with oxygen, when present. Hemoglobin quaternary structure allows it to regulate oxygen activity. When the first oxygen binds, it changes the local structure of the corresponding protein chain, which affects the position of the histidine. The binding of oxygen stimulates movement of the iron atom within the plane of the heme, which pulls the histidine. This movement is propagated to the other chains through their intermolecular interfaces, causing a relatively large motion of the subunits, after which the neighboring chains bind oxygen more easily. Thus, binding the first, second, third, and fourth oxygen is progressively easier, which implies the more oxygen present, the better hemoglobin binds it. Release of oxygen follows the inverse path. Release of the first oxygen is more difficult than release of the second, third, and fourth oxygen; and release of oxygen is easier when little oxygen is present. When the blood is in the lungs, where oxygen is abundant, the protein is loaded with oxygen. In the rest of the body, the lower the amount of free oxygen, the more easily it is released. Hemoglobin picks up oxygen in the lungs and delivers it where needed.

This machinery is so finely tuned that single amino acid changes can lead to major diseases. For example, the replacement of one glutamic acid with a valine in the β-chain causes hemoglobin to form aggregates under

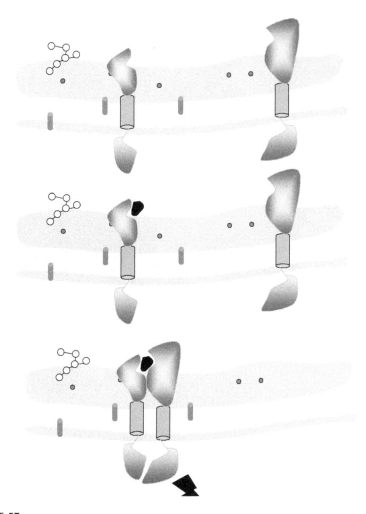

FIGURE 57
The binding of a ligand to the extracellular domain of a transmembrane receptor might cause its binding to a coreceptor (which can be the same or a different protein). The subsequent interaction between the intracellular domains can trigger signaling, for example, by activating a transcription factor that, in turn, activates the required genes.

low-oxygen conditions. This problem causes a distortion of the red blood cells and a loss of elasticity. At the onset of the disease, called sickle cell anemia, red blood cells are capable of regaining their original shape and elasticity when enough oxygen is present. However, with time, the cells permanently lose their elasticity and the ability to flow through narrow capillaries. Although those with the disease die young, the mutation is still frequently found in malaria-infested regions. Individuals who have the mutation in one of their two copies of the hemoglobin gene have a few distorted red cells and are, in general, asymptomatic. However, the parasites that cause malaria spend part of their life cycle in red blood cells, and they

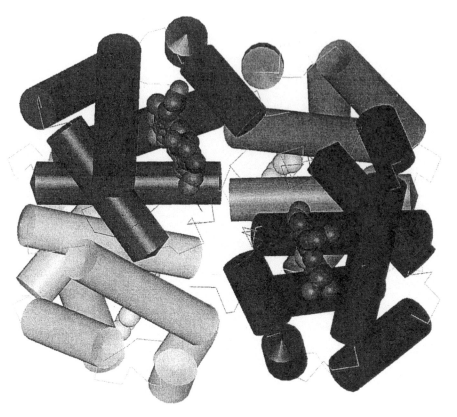

FIGURE 58
Hemoglobin is the oxygen transporter. The protein is α-helical and is formed by four chains, two α-chains and two β-chains bound by noncovalent interactions. It contains one heme group per chain.

cannot survive in cells that contain the mutated hemoglobin. Consequently, carriers of the mutation are relatively less susceptible to malaria.

During pregnancy, the hemoglobin of the fetus takes oxygen from the maternal hemoglobin. The fetal hemoglobin is able to bind oxygen better than its maternal counterpart because the two proteins are different. Fetal hemoglobin is formed by four chains, but the two β-chains are replaced by γ-chains. This chain combination has a higher affinity for oxygen, and, therefore, it can easily accept oxygen from the maternal protein.

What can we learn from the hemoglobin example? Chains can form different interfaces in different conditions, and the same chains (the α-chains) can interact with different chains (the β-chains in adults and the γ-chains in fetuses).

We said that protein chains autonomously fold into their native structure, driven by the enthalpic and entropic gain of the process. However, folding of large proteins with several domains and a quaternary structure can be problematic, and the proteins might not reach their native structure in a

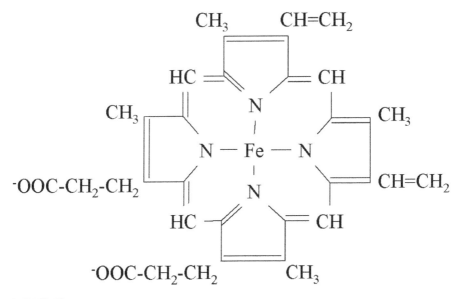

FIGURE 59
The heme molecule. The central iron is bound to four nitrogen atoms, to a histidine side chain (not shown here) and, when present, to an oxygen atom.

reasonable amount of time. What is worse, during the process, they might become stuck in a structure that corresponds to a local energetic minimum and expose large hydrophobic surfaces. This situation is dangerous because it can lead to irreversible aggregation, as is the case of the mutated hemoglobin and of several other proteins responsible for a large number of diseases, such as mad cow disease, Alzheimer disease, and Parkinson disease.

The protein shown in Figure 60 (see color insert after page 40) is used to help the folding process and prevents misfolding. In no way does it alter the final structure that the substrate proteins achieve; it just protects them during the disaster-prone process of folding. Proteins performing this role are called "heat-shock proteins" because their production increases when cells are heated. Heat destabilizes proteins, and when temperature rises, cells need help to fold their proteins.

The GroEL complex is formed by two stacked rings of seven identical protein chains each. The boxes can be closed by a "cap" provided by the GroES protein. The mechanism is simple and, at the same time, elegant. The rings expose a hydrophobic surface at their entrance that binds proteins that expose hydrophobic surfaces and are, therefore, not properly folded. The substrate protein chain enters the box, and GroEL undergoes a conformational change (the energy is provided by the hydrolysis of ATP) that increases the size of the cavity, hides the hydrophobic patch at its entrance, and allows the GroES cap to close it. The misfolded protein is now in a confined, hydrophilic environment, separated by the rest of the cell and in an ideal condition for proper folding. How does the complex "sense" that the process

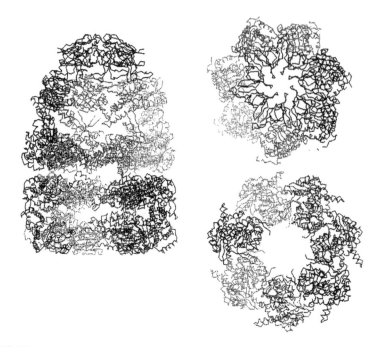

FIGURE 60

GroEL and GroES. The chains of GroEL are shown in different colors. GroES is the "cap" shown in black. The two images on the left are the same protein seen from top and bottom, respectively.

is completed? Apparently, it does not. The opening of the cup and the release of the substrate protein, folded or not, is triggered by another misfolded protein binding to the opposite ring. In other words, the GroEL–GroES complex allows a misfolded protein only until the time when another protein needs assistance.

GroEL and GroES teach us that proteins can use conformational changes to trigger binding and release of other components, which allows the composition of a complex to be regulated by external conditions.

Viruses exploit the biological machinery of their host cells and force them to make new viruses, often killing the cells in the process. Picornaviruses (little RNA viruses) are formed by protein shells that enclose RNA. The viruses that cause polio and the common cold are picornaviruses.

The shells of picornaviruses have an icosahedral shape and are composed of 12 identical pentagonal substructures. In polioviruses and rhinoviruses, each of the icosahedral "pentagons" is made of three different proteins. A fourth protein is in the interior of the shell and ensures that the pentagonal units are properly oriented. The complex has to be stable to survive in the environment (rhinovirus can survive for days on our hands and still be infectious), but they must also be able to disassemble to release the nucleic acid once they have entered the host cell.

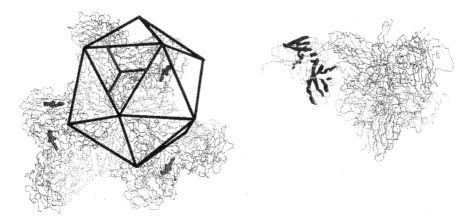

FIGURE 61
The structure of a portion of the rhinovirus coat bound to an inhibitor (left) and to an antibody (right). The virus has an icosahedral shape. Each of the triangular faces is formed by three protein chains shown in different colors.

Vaccination is a defense against viral infections. It is a process by which one artificially teaches (primes) an immune system to recognize a virus by injecting a portion of an inactivated form of the virus in an organism. The injection elicits the production of antibodies that bind to the virus and clear the infection. For viruses that mutate often, such as rhinovirus, vaccination is less effective, and one has to resort to drug therapy. Some drugs act by blocking the sites at which viruses gain entrance to cells. Other drugs are designed to stabilize the viral structure, which makes the virus unable to open and deliver its nucleic acid.

Knowledge of the quaternary structure of the target virus is essential for both drug design and the development of vaccines. Figure 61 (see color insert after page 40) shows the structure of a portion of the rhinovirus coat with a bound drug molecule and the complex between a rhinovirus particle and an antibody fragment.

Antibodies are our first defense line against infections. They circulate in the blood and bind to the surface of foreign molecules. In some cases, such as in rhinovirus or poliovirus infections, the action of antibodies can be enough to block infection. In other cases, such as in bacterial infections, binding of antibodies constitutes a signal that triggers other defensive mechanisms of the immune system. Antibodies are paradigmatic examples of proteins that form transient complexes. Although stable by themselves, they are able to form very tight complexes with foreign molecules.

The human body contains up to 100,000,000 different types of antibodies, and each recognizes a different target molecule. Antibodies are pre-existent to the infection, and their ability to bind almost any foreign molecule is the result of their structural properties and their amazing variability. So many different kinds of antibodies are available that the probability that at least some are the right ones to fight an infection is very high. Each antibody is

FIGURE 62

The structure of a fragment of an antibody (PDB id: 3HFL) bound to an antigen. The contact region of the antibody is shown in green dots, and that of the antigen (lysozyme) is shown by a solid blue surface.

formed by two heavy chains and two light chains, which comprise about 450 amino acids and about 220 amino acids, respectively (Figure 62) (see color insert after page 40).

If we have this many antibodies, do we not need billions of nucleotides to code for them? In reality, this big collection is created by recombination (rearrangement) of a relatively small number of genes in lymphocytes, the blood cells that make antibodies. Each lymphocyte recombines its antibody genes in a different way and makes a different type of antibody. If a specific antibody binds a foreign agent (antigen), the lymphocyte that produced it multiplies, and more antibodies that are able to recognize the foreign molecule are produced. In the daughter cells, the antibody sequences undergo some small changes, thus increasing the probability that some of them will bind the antigen more tightly and specifically.

Antibodies are one of the best examples of how much can be learned about a protein by comparing its sequence with other members of the same class. The antibody sequence contains regions with similar sequences, which suggests that they correspond to domains that share a similar fold, two in the light chains and four in the heavy chain. One of the four domains of the heavy chain and one of the two domains of the light chain are relatively more variable among different antibodies than the others, which suggests that these variable domains are deputed to binding the antigen. Within each variable domain, three regions show an even higher variability, possibly because they are exposed to the solvent and directly involved in antigen binding. All these earlier observations based on sequence analysis were confirmed upon determination of the structure of the first antibody.

How can a single architecture be used to bind so many different molecules? Each antibody has a different sequence obtained by genetic recombination of several segments. How does nature avoid having most of the combinations result in unfolded proteins? An antibody is formed by β-sheets packed against each other, and the antigen-binding regions correspond to loops that connect the strands. These loops can only assume a limited set of conformations (called canonical structures) of their main chain. The specific conformation depends upon the presence of a few specific residues in key positions; all other amino acids are free to vary and generate surfaces with different topographies (Figure 63). In this way, by maintaining only a few fixed main chain–determining residues, antibodies can generate different binding surfaces, which minimizes the risk of destabilizing their overall structure. Indeed, antibodies are the only proteins for which we can predict the conformation of loops on the basis of their sequence alone. We only need to know which ones are the key residues to predict the main-chain structure of the loop. This finding holds for five of the six loops and for a portion of the sixth.

The impressive variety of binding modes of antibodies and their limited possibility of changing the main-chain structure of the binding regions teaches a very important lesson: recognition is mainly mediated by side chains. This statement is not good news for us, because exposed side chains are very flexible, and, therefore, their conformation is difficult to predict. This observation has several implications for the prediction of macromolecular complexes.

Sequence-Based Methods for Predicting Interactions

The number of available protein sequences is by far larger than the number of known structures. Therefore, the ability to detect which proteins interact

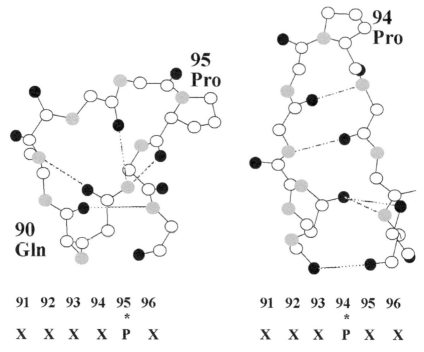

FIGURE 63

The canonical structures of immunoglobulins. The loop shown is called L3 (it is the third loop of the light [L] chain of antibodies and is part of the antigen-binding site). When the length of the loop is six amino acids, as in the figure, only two main-chain conformations are observed. The one on the left occurs when the amino acid in position 95 is a proline and the amino acid in position 90 is a glutamine. The conformation shown on the right occurs when the proline is in position 94. All other residues are free to vary and contribute to shape the antigen-binding region.

with each other, and with which regions, from their amino acid sequences alone would be extraordinarily more useful than a structure-based method. How can we achieve this goal? How can the amino acid sequence of a protein be used to infer which, if any, of the tens of thousands of other proteins it will bind to?

One possibility is to search for cases in which two interacting proteins are separated in one species but are present as two domains of the same protein in another species. This occurrence is a strong indication that the two proteins interact, either physically or logically. This method has a very high accuracy, but, as expected, its coverage is not very high. In other words, when we find such a case, confidence is high that it is indicative of a true interaction, but the number of such cases we can find is not very high.

Another possibility is to look at comparative genome analyses. Now that many genomes are available, we can try to detect the proteins that co-occur in different genomes. Two proteins that are either always both present or always both absent are likely to interact with each other. However, with this method, we would not be able to discriminate between a physical and a logical interaction. One protein might, for example, modulate the production of the other by binding to regions of the DNA that control the expression of its gene.

Conserved proximity of two genes in genomes has been also used to detect proteins likely to be part of the same biological process. Unfortunately, the correlation between genomic localization and function is high in prokaryotes (simple-celled organisms that lack a defined nucleus) but much less strong in higher organisms.

Another strategy is based on the consideration that the evolution of different protein families might be dissimilar. When we use sequence similarity to build phylogenetic trees for different protein families, the length of the branches reflects the rate of evolution. If we assume that the rate is constant, then we can associate it with evolutionary time. However, different proteins can be subjected to different evolutionary pressures and, therefore, might appear as more or less recently diverged with respect to other proteins in the same organisms. If two protein families show the same rate of evolution (i.e., if the trees built on the basis of their sequences are similar), this similarity might be the result of coevolution brought about by the presence of a physical interaction between the two proteins. Similar conclusions can be derived from searching scientific publications for statistically significant co-occurrences of two protein names.

If we know, or have been able to predict with some reliability, that two proteins form a complex, we need to know where they interact; that is, which residues of one interact with which residues of the other. We can use the structures or models of the two proteins and look for geometrical complementarity or energetically favorable interactions. Another strategy, also used for the prediction of structural features of single proteins, is correlated mutations. In this method, we analyze the variations of each of the positions in a multiple-sequence alignment to detect positions that change in a correlated fashion. For example, if two positions are always occupied by charged amino acids, different in different sequences but always of opposite charge, we can hypothesize that they are in contact in the structure. Similarly, if two hydrophobic residues are in two different positions, and every time one of them is relatively big the other is small and vice versa, the possibility is strong that this circumstance is caused by the fact that they are close in space and have to fill an equivalent hydrophobic pocket in the proteins. Only rarely can these methods reliably predict a large set of interactions, but sometimes they can provide a few sufficiently reliable pairs, and these pairs can be used to discriminate between different alternative folds for a protein sequence.

If we know that two proteins interact, and we have many sequences of members of their evolutionary family, we can detect correlations between positions in the two alignments. This idea is impaired by the (presently) low probability of having large families of two proteins spanning the same species so as to allow detection of correlated mutations by use of corresponding pairs.

All these methods have been used to derive maps of putative protein interaction networks in organisms. In general, the results are intricate, hard to understand graphs in which each protein is associated to a node and each predicted interaction is depicted as an edge that connects the interacting proteins. This static representation is far from being realistic because some interactions can be mutually exclusive. A protein might bind either one or the other partner in different conditions in the same physical locus, and, therefore, the two interactions cannot occur at the same time for the same molecules. This problem is also common to interaction maps drawn on the basis of experimental results.

How do we evaluate the efficacy of methods for predicting interactions? Only a small set of proteins have been biochemically characterized, and we need large experimental data sets to validate our methods. Far from being purely theoretical, this question is the central point to be addressed. The genomic era made possible high-throughput experimental techniques for detecting interactions, but their accuracy is not very satisfactory at present, and this inadequacy impairs our ability to test and improve the computational methods.

Experimental Methods for Detecting Protein–Protein Interactions

Affinity chromatography is a classical method of detecting whether a protein interacting with a ligand is present in a mixture of proteins. The ligand is immobilized on a matrix, and the mixture is added to this "functionalized" matrix. A protein of the mixture is retained on the matrix if it interacts with the ligand. After eliminating the unbound components of the mixture by washing, the bound proteins can be eluted (e.g., by adding an excess of the ligand) and characterized (Figure 64).

This methodology can be tailored to the detection of which proteins present in a cell bind to a protein of interest. The protein is cloned and its gene is fused with the gene of, for example, glutathione S-transferase or LacZ, two proteins that tightly bind to a molecule called glutathione or to a region of DNA that codes for a stretch of histidines that bind zinc. The fusion protein binds to immobilized glutathione or zinc through the artificially added binding domain. At this stage, a protein extract from the cells under investigation is made to interact with the matrix, and unbound proteins are washed away. Proteins that bind to the protein of interest are retained and can be subsequently eluted and characterized by, for example, mass spectrometry.

Another technique is the so-called yeast two-hybrid system. Yeast has a protein called gal4 formed by two domains. The first domain is responsible for binding specific DNA sequences upstream of the galactosidase gene. The

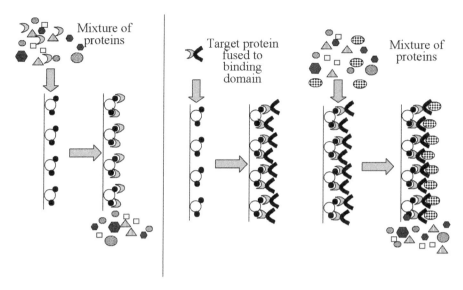

FIGURE 64

Scheme of an affinity chromatography experiment. The scheme on the left can be used to select a protein binding to an immobilized ligand. On the right, the target protein is fused to a ligand-binding domain through which it binds the matrix. A protein interacting with it can be isolated from a mixture of other proteins.

second domain is responsible for activating the transcription of the gene. The system works only if the two domains are physically close. If the gene coding for gal4 is split into two parts, each of which codes for one of the domains, no activation of transcription occurs. The gal4 domains can each be fused to two different proteins. If the proteins form a complex, their interaction brings the two gal4 domains close in space, and transcription of the downstream gene is activated. The galactosidase gene can be replaced by a gene that confers the ability to survive in conditions that are specifically lethal to yeast, so that a cell survives only if the two proteins under investigation interact with each other (Figure 65) (see color insert after page 40).

A high-throughput version of this assay consists of constructing one yeast plasmid (i.e., a small, circular DNA molecule, separate from the bacterial chromosome, capable of independent replication) that contains the gal4 DNA-binding domain fused to the protein of interest (bait) and a library of yeast plasmids, each of which contains a different gene fused to the transcription-activation domain gene. The first plasmid is inserted (more correctly, "transfected") in each cell of a yeast population. The same population is also transfected with the library of plasmids that contain genes fused to the transcription-activation domain gene. Only the cells transfected with a gene whose product interacts with the bait will survive. The plasmids contained in these yeast cells can be sequenced to identify the interacting proteins.

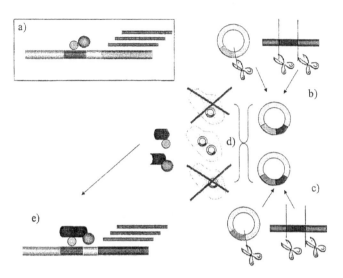

FIGURE 65

A simplified view of the rationale of a yeast two-hybrid experiment. The gal4 system is shown in (a). The gal4 protein, composed of a DNA-binding domain and an activation domain, activates the transcription of the downstream gene. Two genes coding for two interacting proteins are cloned and fused to each of these domains as shown in (b) and (c). The plasmids are transfected into a yeast cell population. Only in cells that contain both fusion proteins, the spatial proximity between the DNA-binding domain and activation domain is reconstituted, which leads to transcription of the downstream gene. If the original gene is replaced by a gene essential for cell survival, only cells that contain both fusion proteins can survive. This system can, therefore, be used to determine whether two proteins interact. Moreover, if the protein of interest is fused to one of the domains and a population of proteins is fused to the other domain, the method allows the identification of which proteins in the population, if any, interact with the protein of interest.

Phage library display is a method of detecting interactions that is also suitable for identifying amino acid sequences able to interact with a ligand (see Problem 10). Another technique is chromatin immunoprecipitation (ChIP), which is used to detect DNA sequences recognized by particular proteins, such as transcription factors.

All these techniques detect interactions of different strength, but they all have drawbacks and problems, of which the most serious is the fact that the ability of two proteins, or a protein and a nucleic acid fragment, to physically interact does not necessarily mean that they do so physiologically. They might never meet each other, because they are in different compartments, or because the concentration of one or both of them in vivo is not high enough for them to form a complex. Furthermore, previously known interactions are not all detected by these techniques. Not surprisingly, the overlap between the interactions detected by different experimental and computational techniques is rather small, and we still face the serious problem of how to validate them.

Structure-Based Methods for Predicting Interactions

The problem of protein docking involves finding the coordinates of the complex of two molecules, given the coordinates of each molecule separately. We must make a distinction between what we call "bound" and "unbound" docking. The distinction is based on whether we use as input the coordinates of the two molecules as they are in the bound complex or in the unbound complex. In other words, we can reconstruct the relative position of two molecules in a complex starting from the experimentally determined coordinates of the components of the complex, or we can start from the experimental or modeled structure of the two unbound proteins. Clearly, the former is an exercise, whereas the latter is a biologically relevant problem. However, the latter is also a much more difficult challenge because of the aforementioned problem of flexibility. In the "unbound" docking experiment, the position of the exposed side chains, and, therefore, the shape of the surface, is not the same as in the complex. Thus, the detection of complementary surfaces is not enough. We must also search for potentially complementary surfaces, take into account how the structure of the two interacting partners is affected by binding, and evaluate the likelihood of the presence of water molecules buried in the interface between the two proteins. The only reason the results of "bound" docking are often presented for a given method is that the number of cases in which we know both the structures of the unbound components and the structure of their complex is rather low.

Successful "docking" depends on three things: how we represent the protein structures, taking into account their flexibility; how we search the conformational space of the possible solutions; and how we evaluate and rank the solutions.

Representation of Protein Structures for Docking

The methods we used for surface description in the context of finding similar surfaces in different proteins are also applicable to the docking problem. As illustrated in Figure 66, a protein surface can be represented by its normal vectors, by a set of spheres, by equally spaced vectors radiating from the center of the molecule, or by embedded vectors in a three-dimensional grid.

Computational Approaches to Include Protein Flexibility in Docking Procedures

In principle, any algorithm that is applicable to surface structural comparison can be applied to docking and vice versa. However, if we are attempting to

Radial projection: each point on the sphere contains a feature of the respective protein point (for example its distance from the center)

Sphere with the center on the surface normal and a radius such that it touches a surface point but does not intersect the surface

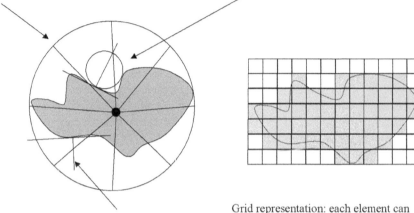

Surface normal

Grid representation: each element can contain a logical value (occupied or free), but it can also contain values related to the properties of the grid elements (hydrophobicity, charge, evolutionary conservation, etc.)

FIGURE 66
Examples of representation of protein surfaces (in two dimensions).

detect common arrangements of functional residues, we can hypothesize that their conformation is well-conserved. Conversely, in docking, flexibility plays an important role. A protein structure, and especially the side chains involved in binding, can undergo quite substantial rearrangements upon formation of the complex. Furthermore, buried water molecules are often found in the interfaces.

One possible approach is to generate many different conformations of the two target proteins and use each of them as input of a docking procedure. The different conformations can be a set of different structures determined by X-ray crystallography in different conditions, or the structures can be generated by an NMR experiment. (The result of an NMR experiment is, by and large, a set of intramolecular distances that are used to derive a set of three-dimensional structures that satisfy the measured distance constraints.) We can also generate an ensemble of different structures for the two proteins by running a molecular-dynamics experiment and collecting structures during the simulation, or we can use genetic algorithms or Monte Carlo simulations. These simulations can be tailored to our problem if we only allow motion or variation of exposed side chains.

The more numerous our structure set, the more likely that one of the structures resembles the bound conformation of our target protein. However, the calculation is more expensive, and the likelihood of each of our long list of putative solutions is more difficult to evaluate. The search procedure for docking two molecules is computationally intensive, and it cannot be applied to large sets of pairs of initial structures. Some methods superimpose the structures of the initial conformational ensemble and define an average structure in which each atom is assigned a variability dependent upon how different its position is in the various members of the ensemble.

Side-chain rotamer libraries are compilations of the most observed side chain conformations for each amino acid chain, and they can be general or backbone-specific; that is, we can derive a frequency table for each possible conformation of amino acids collectively or separate each one on the basis of the type of secondary structure in which it is observed. Some docking methods take advantage of these compilations and assign a variability to each amino side chain on the basis of the repertoire of its side-chain rotamers. In some cases, the observation of the protein structure might suggest specific regions that are likely to modify their conformation upon binding. For example, regions between domains can be specifically allowed to assume a discrete set of conformations.

Another possibility is to use a single-protein structure conformation and introduce the treatment of flexibility later in the procedure, by not penalizing limited overlap between the two proteins' atoms; that is, to treat each atom as a soft rather than a hard sphere so that limited interpenetration of the atoms of the two proteins is allowed in the final docked conformation.

Searching Conformational Space for Docking

Protein docking is one of the most imaginative research areas in protein bioinformatics, and enumerating all the methods and algorithms that have been tried is very difficult. In all cases, however, the procedure is computationally intensive. We are looking for patches of the proteins' surface that match each other, so we must explore the complementarity of every pair of surface patches of reasonable size in the two proteins for each rotation and translation of one of the proteins with respect to the other. To this end, we must estimate the average dimension of complementary patches; that is, the interaction surface in a complex. Given two interacting proteins, A and B, the strength of their interaction can be measured by their dissociation constant K_d, defined by

$$[A][B]/[AB] = 1/K_d$$

where [A], [B] and [AB] are the concentration of the two proteins and of their complex at the equilibrium, respectively.

In known protein complexes, the value of K_d ranges from 10^{-4} to 10^{-14}. The strength of interaction is very variable, and this characteristic is reflected by the physical parameters of the complex. For example, the surface of

interaction, defined as the decrease in solvent accessible surface upon binding, can vary between 1,000 and 5,000 Å^2 and represent 6% to 30% of the
total protein surface. The majority of the interface areas for complexes of
known structure measure around 1,600 Å^2 ±400 Å^2, and most algorithms
are tailored for searching complementary regions of this size, but the size
can be as small as 1,000 Å^2.

To detect these interacting regions, we can either perform a full search of
the conformational space, which involves trying every possible relative orientation of the two proteins, or we can use stochastic methods such as Monte
Carlo, simulated annealing, molecular dynamics, and genetic algorithms,
which can guide the search toward the best solution.

Exhaustive search of the space is computationally very expensive. Even if
we assume that the two molecules are rigid, we still must explore six degrees
of freedom, three for the translation and three for the rotation of one molecule
with respect to the other. Thus, the number of comparisons can be as high
as several billion. Such was the case for one experiment that successfully
docked the electron transfer complex between bovine cytochrome *c* oxidase
and horse cytochrome *c* (more than 30 billion configurations were tested and
scored). In that case, no alternative conformation of the two protein structures was used. In other cases, in which binding induces larger movements,
even such a high number of trials might not be sufficient.

The grid representation of two putatively interacting molecules can be
Fourier transformed so that the translation operations can be performed in
Fourier space by convoluting the Fourier spectra. This process saves considerable computer time. This approach is shown in Figure 67. The computational time depends upon the number of grid points and on the angle by
which the second molecule is rotated at each iteration. A typical value of the
latter is around 15°. Therefore, the cycle has to be repeated a few thousand
times.

If information on the rough location of the binding surface is available
(e.g., we know where the antigen binds in the case of antibodies), the search
space can be reduced. However, when we discuss the results of the CAPRI
experiment, a blind assessment of methods for protein docking, we will see
that sometimes our assumptions might be misleading.

Scoring Docking Solutions

Although docking procedures differ widely, they all generate a large number
of putative solutions, generally ranked according to some approximate
energy-based score. The correct solution, if present in the list, might rank
among the top 10 or 100, according to the method used, and false positives
are a serious problem in each method. As it is often the case, nature does
not behave in an easily predictable fashion. The "true" answer does not
necessarily have the largest interaction surface or the more hydrophobic one.
It does not always contain the highest number of hydrogen bonds or the

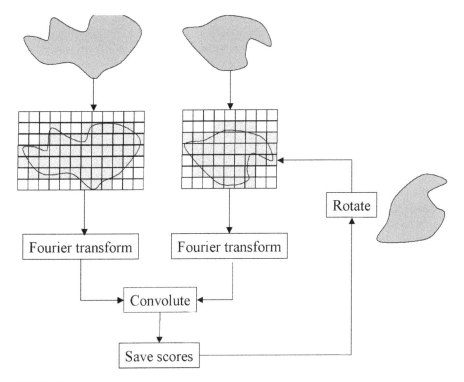

FIGURE 67
Schematic representation of the application of Fourier transforms to docking. The cycle is repeated until the rotational space of one of the two molecules has been sampled. The level of detail depends upon the size of the grid spacing and of the rotational angle.

smallest number of buried polar groups. The problem of detecting which among the many solutions is the correct one is, therefore, far from trivial.

Available methods often use a two-stage ranking. The first stage, with an approximate and fast-to-compute function, is used to eliminate very unlikely solutions. The second, more accurate, stage is then applied to select the best among the remaining solutions.

A combination of several parameters is commonly used. Some parameters contribute positively to the score, whereas others decrease it (penalizing the solutions). In some methods, the presence of many similar solutions is taken as an indication of correctness, so investigators cluster all their solutions and use the size of each cluster as one of the scoring parameters.

The analysis of known protein complexes reveals that the surfaces of the two components are a nearly perfect match, so measures of geometric complementarity are virtually always present among the scoring parameters. For example, we can check for nearly opposite surface normals or for matches between edges of a graph that represents the surfaces. Surface complementarity is very effective in selecting the correct solution in "bound" docking experiments, but in real cases, the flexibility of the protein surface and the putative presence of water molecules at the interface limit their accuracy in discriminating between different solutions. To increase the effectiveness of scoring on the basis of surface complementarity, we should tolerate some clashes, or interpenetration, of surfaces, the extent of which depends on the type of side chains. A greater extent of overlap can be allowed for large flexible amino acids such as lysine, arginine, aspartic acid, glutamic acid, and methionine. In some cases, the allowed tolerance is calculated on the basis of an experimental measure, the B factor. This measure is a parameter calculated for each atom of a protein structure determined by X-ray crystallography and is related to the mobility of the atom in the protein.

Another parameter that is included in most scoring functions is a measure of the number of hydrogen bonds formed between the two protein components. Often, methods also take into account the number of atoms that could form hydrogen bonds but do not. An atom that does not form a hydrogen bond in the complex but does so when the protein component is free in solution negatively contributes to the free energy of the complex. Given a putative docking solution, we should compute the extent of the contact area composed by hydrophobic atoms, as well as the percentage of cases in which a hydrophilic residue is in contact with a hydrophobic residue and vice versa.

Known protein complexes can be used to calculate the expected frequency of each pairwise interaction, either atom based or residue based, and to derive an approximate free-energy charge ΔG for their interaction, similar to what we described for the prediction of protein structure.

Electrostatics and solvation are two important scoring parameters. They are usually taken into account by use of some rather crude approximations. However, these effects are not simply related to Coulomb interactions between charged atom pairs of the two interacting proteins or to the loss of hydrogen bonds between water and the interacting proteins, as is shown in Figure 68.

The CAPRI Experiment

How well do the many methods of predicting interactions work? The Critical Assessment of Predicted Interactions (CAPRI) experiment is designed to answer this question. CAPRI is a worldwide effort modeled on the CASP experiment described in Problem 4. Its purpose is to evaluate predictions for the structure of a soon-to-be-known complex of two proteins, the structures of which have already been solved independently. Not surprisingly, targets for the CAPRI experiment are quite difficult to find. Therefore, targets are collected and predicted when they become available, rather than at specific, predetermined periods, as in CASP, and the assessment is carried out when a sufficient number of targets (and predictions) has been collected.

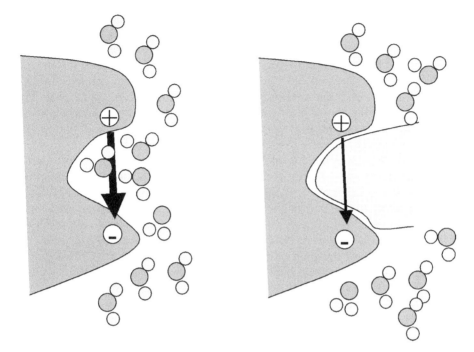

FIGURE 68

The energy of the electrostatic interaction between the two opposed charges in the darker protein can be modulated by docking of another protein. The dieletric constant for the conformation shown on the left is roughly 80, as the space between the two charges is mainly filled by water. On the right, the dieletric constant depends upon the atom distribution of the interacting protein. At the same time, bound water molecules are released into the bulk solvent, which causes them to gain entropy.

To increase the set of possible targets, in some cases, the structure of one of the two proteins is taken from the complex. These cases are regarded as intermediate between a "bound" and "unbound" docking. Nevertheless, the number of targets is never very large, and an accurate evaluation of the state of the art, as well as an estimate of how much progress is made with respect to previous editions, is still difficult.

A prediction is considered successful if the rmsd between the predicted position of the atoms at the interface and their real position in the complex is lower than 4 Å and if more than 10% of interacting amino acid pairs in contact have been identified. These thresholds can seem rather generous, but in the first and second edition of CAPRI, no group correctly identified more than 33% of the interacting surface pairs.

The conclusions that can be derived from CAPRI, because of the limited number of targets, should be taken with even more care than other statistical assessments of methods. However, in 60% of the cases, at least one docking method was able to correctly identify the binding site, and different search algorithms and score functions achieved a similar rate of success (or failure).

The results of CAPRI are instructive. Camelids have the classic mammalian antibodies, formed by light and heavy chains, but they also produce antibodies formed only by a heavy chain. The complexes of one of these antibodies with three different ligands were targets in CAPRI and, for two of them, the binding region did not include only the classical antigen-binding regions. None of the methods identified the correct solution, probably because all predictors assumed that they would behave as regular antibodies. This failure shows that, in some cases, predefinition of the binding site can be seriously misleading.

Another target was the complex between a protein kinase (an enzyme that phosphorylates other proteins) and its substrate. Both the site of phosphorylation of the substrate and the location of the active site of the enzyme were experimentally known, so this case should have been an easy one. However, the kinase undergoes a substantial conformational change upon binding (rmsd between the bound and unbound form was 2 Å), and, therefore, the quality of the predictions was not very satisfactory. Only 8 out of 63 predictions (each group can submit more than one prediction) identified more than 10% of the correct residue pairs and had an rmsd of the interface atoms lower than 10 Å. The subsequently determined free structure of a protein homologous to the kinase is more similar to the bound form of the target enzyme than that of its unbound form. This finding might suggest that an analysis of structural divergence between proteins of the same family can be useful to model flexibility.

Promising Avenues

The two routes for detecting sequenced-based and structure-based interaction would both benefit from a larger data set of examples and from a more reliable data set of experimental results. These data sets will most likely become available in the near future. High-throughput methods are bound to achieve a better accuracy, and this improvement will, by itself, allow computational methods to achieve a higher reliability.

Clearly, a major problem is how to treat flexibility. The continuous increase in computer power might help, but understanding the basic principles that govern the motion of a protein upon binding its partners is also important. The large number of structures that are becoming available, and the advances in protein structure prediction methods, will certainly contribute to our understanding.

Whereas structure-based methods have been around for some time, sequence-based methods are rather recent because we only now have a sufficiently larger set of examples. A test to determine whether the combination of the two approaches can lead to better results would be useful. The docking problem is complex, and we must exploit all available information, in terms of both sequence and structure data.

Suggested Reading

Halperin, I., Ma, B., Wolfson, H., and Nussinov, R. Principles of docking: an overview of search algorithms and a guide to scoring functions, *Proteins* 47, 409–443, 2002.

Janin, J. and Chothia, C. The structure of protein-protein recognition sites, *J. Biol. Chem.* 265, 16027–16030, 1990.

Perutz, M.F. Stereochemical mechanism of oxygen transport by haemoglobin, *Proc. R. Soc. Lond. B Biol. Sci.* 208, 135–162, 1980.

Gobel, U., Sander, C., Schneider, R., and Valencia, A. Correlated mutations and residue contacts in proteins, *Proteins* 18, 309–317, 1994.

Chothia, C., Lesk, A., Tramontano, A., Levitt, M., Smith-Gill, S., Air, G., Sheriff, S., Padlan, E., Davies, D., Tulip, W., Colman, P. M., Alzari, P. M., and Poljak, R. J. Conformations of immunoglobulin hypervariable regions, *Nature* 342, 877–883, 1989.

Janin, J., Henrick, K., Moult, J., Eyck, L.T., Sternberg, M.J., Vajda, S., Vakser, I., and Wodak, S.J. CAPRI: a critical assessment of predicted interactions, *Proteins* 52, 2–9, 2003.

Smith, G.P. and Petrenko, V.A. Phage display, *Chem. Rev.* 97, 391–410, 1997.

Bartel, P.L. and Fields, S. *The Yeast Two-Hybrid System*, Oxford University Press, Oxford, UK, 1997.

Pandey, A. and Mann, M. Proteomics to study genes and genomes, *Nature* 405, 837–846, 2000.

Kirmizis A. and Farnham, P.J. Genomic approaches that aid in the identification of transcription factor target genes, *Exp. Biol. Med. (Maywood)* 229, 705–721, 2004.

Problem 8

Protein–Small Molecule Interaction

Introduction to the Problem

The number of algorithms available for protein–protein docking is large, but the number of possible methods for docking a small ligand and a protein is even larger. This observation is not surprising, given the centrality of this problem to many biological processes and, even more so, to pharmacological applications.

Docking of a small ligand to a protein receptor is performed to answer the question of where the molecule binds, in which orientation, and with which affinity. When successful, this docking can give invaluable information about how to modify the molecule to increase its affinity and its specificity. This process is related to the problem of designing a ligand for a given receptor (de novo ligand design). The methods are not very different and are not discussed separately here.

Search Strategies and Scoring Functions

As in the case of protein–protein interaction, we need a search strategy to explore the various modes of binding of a small molecule to the target protein and a scoring function to assess the likelihood of the proposed docking solution. Exhaustive search of the complete space available to a ligand around a protein of average size is computationally impractical here, as in the cases described in the previous problem. Early methods treated both the protein and the ligand as rigid bodies and did not allow any flexibility of either molecule. However, the limitations introduced by treating the protein as a rigid body are as serious as those caused by the treatment of a small ligand as rigid. Although some methods still consider the receptor as rigid, all widely used methods take into account the flexibility of the ligand.

Commonly used techniques for small molecule docking are based on methods we have already described, such as molecular dynamics, Monte Carlo, and genetic algorithms. Other domain-specific methods include fragment-based methods, point complementarity, distance geometry, and systematic searches. These methods are often combined in a two-stage approach in which a fast, computationally inexpensive method is used for prescreening of ligand positions that are later optimized by a more computationally extensive method.

Fragment-Based and Point-Complementarity Methods

Fragment-based methods select a "relevant" fragment of the ligand, exhaustively search for its best docking position, and grow the rest of the molecules around it. The choice of the base fragments is clearly very important in these techniques. Ideally, they should represent important functional groups with limited internal degrees of freedom, so that their flexibility is less of an issue. A modification of this procedure is widely used in the design of new ligands. We can select a subset of important functional groups, optimize their position in the selected binding region, and subsequently try to design a molecule that contains them in the correct relative position.

A very commonly used program based on this rationale is FlexX. Its first step is the selection of one base fragment, either manually or automatically. The program searches for the optimal relative position with respect to the receptor by fitting three sites of the fragment to three sites of the receptor. The fit takes into account hydrogen bonds and hydrophobicity, and the fragment is considered rigid. All placements of the ligand are clustered and scored by application of an energy-based function. The next fragment is added to the optimized base fragment in all possible positions and conformations. Cases of intramolecular or intermolecular overlap are removed. The best solution for each step is used in the next iteration until the complete ligand is built. Finally, the resulting solutions are scored. Recent modifications of the algorithm precalculate putative favorable locations of water molecules in the binding site that are then included in the calculation.

Another very popular program, LUDI, uses a three-stage protocol. In the first stage, it calculates the positions in space that can make favorable interactions with the receptor; for example, the ideal location for a hydrogen donor group or for a hydrophobic atom. In the second stage, fragments of the ligand are fitted to the favorable interaction points. In the third stage, fragments are joined, with a protocol similar to that described for FlexX. The favorable sites of interactions are computed by one of these three strategies: (1) on the basis of a statistical analysis of a database of small molecule structures (the most used small molecule database is called Cambridge Crystallographic Database), (2) from geometrical considerations, or (3) from the output of the program GRID. The latter program embeds the receptor into a three-dimensional grid and calculates the binding energy for a set of probes (e.g., amide, aromatic carbon, and oxygen) located at each point of the grid. Some methods, such as FLOG, use similar approaches but consider more than one conformation of each fragment to address the flexibility problem.

Distance Geometry-Based Methods

Distance geometry was originally developed to solve experimental NMR structures. As we mentioned, the result of an NMR experiment includes a set of distance constraints between pairs of atoms. The objective of distance geometry is to find a solution that satisfies as many distance constraints as possible.

If we have a molecule composed of N atoms, the number of distances is $N*(N-1)/2$, and the number of coordinates is $3N$. Therefore, if we had all the distances between every pair of atoms, and they had infinite precision, we could reconstruct the structure. The problem is to reconstruct the three-dimensional object on the basis of an incomplete and approximate set of distances. The basic idea is to construct the matrix $g_{ij} = x_i x_j$; that is, the matrix that contains the scalar products of the vectors x_i representing the atom coordinates. The element of the matrix can be written as

$$g_{ij} = x_i x_j = \frac{1}{2}(d^2_{iO} + d^2_{jO} - d^2_{ij})$$

If the origin O is chosen as the centroid of all the atoms, the distances from the center can be computed from the distances alone. This implies that we can calculate the values of g_{ij} and, consequently, the coordinates of x_{ij}. The mathematics are beyond the scope of this text. Therefore we mention only that the solution is

$$x_{ik} = \lambda_k^{1/2} \, w_{ik}$$

where λ_k and w_{ik} are the eigenvectors and eigenvalues of the matrix g. For the distances to correspond to a three-dimensional object, only the first three eigenvalues of the matrix should be positive and all the others should be equal to 0.

If the input distances are not exact, more than three eigenvalues will be different from 0, but the method can still be applied by use of the three largest eigenvalues. If the matrix is incomplete, we can approximate the unfilled values by use of the three-dimensional inequalities ($d_{AC} < d_{AB} + d_{BC}$ and $d_{AC} > |d_{AB}-d_{BC}|$).

Distance geometry can be applied to docking problems by selecting different sets of distances between, for example, potential hydrogen bond donors and acceptors of the two molecules. In one approach (DockIT), the binding site is represented as a set of spheres, and several ligand conformations are generated within the site by use of distance geometry.

Virtual Screening

Every large pharmaceutical company has a huge collection of compounds, synthesized for different projects, bought from chemistry labs, collected as reaction intermediates in synthetic processes, or derived from natural sources. In a typical drug discovery process, each of these compounds is screened for activity against the selected target molecule or process so that potentially active molecules can be selected. These molecules are subsequently characterized and optimized, if endowed with favorable properties. The process is usually automated, and, if the activity assay for screening can be properly designed, millions of compounds can be tested in a matter of weeks. Nevertheless, some interest exists in preselecting a set of molecules with a higher probability of having a high activity against the selected target and, possibly, a low binding affinity for other similar molecules. The setup of a suitable experimental screening assay that can be automated and run in high-throughput mode is not necessarily straightforward and, in some cases, the only assay available is time- and labor-consuming. Furthermore, smaller enterprises have limited access to large compound collections.

The development of fast docking methods and the parallel evolution of computers has created the possibility of testing large collections of compounds *in silico*, with the objective of reducing their number to a manageable subset. This process is called virtual screening, and most of the methods previously described have been or can be used in virtual screening applications.

The problem is ideally suited for parallelization, as each compound can be screened and evaluated independently of any other on the target structure. A recent development is to use distributed computing, which takes advantage of a technology originally developed for searching for extraterrestrial life. Desktop computers are provided with a program, a target protein, and a set of compounds. Whenever the computer is idling, its CPU is used to run the docking program. This strategy has been applied to a virtual screening experiment designed to find inhibitors of a protein of the anthrax virus. In less than a month, 3.5 billion compounds were tested by more than 1.5 million computers in more than 200 countries. Computer power is not a problem anymore for problems that are intrinsically parallel.

The Properties of a Drug

Docking methods can find molecules that bind a given receptor and, in many cases, block its activity. Binding to the appropriate target receptor with a good affinity is but a first step toward the development of a drug and probably the least important one for pharmaceutical companies, as any failure in the development of a drug is more expensive the later it occurs during the development process. Identifying which properties of a compound make it a "good" drug is of enormous interest.

Certainly, a drug must have a high affinity for the receptor. This feature decreases the necessary dosage and reduces the probability of spurious binding to other molecules and the occurrence of side effects. However, other very important parameters are its synthetic accessibility, its lack of reactivity, oral bioavailability, favorable pharmacokinetics, and appropriate elimination pathways.

Oral availability correlates with low molecular weight (it should be lower than 500 Da), low number of hydrogen bond donors (fewer than 5) and acceptors (fewer than 10), and low lipophilicity. For other properties much less is known, but we have a set of drugs in the market, and they represent solved examples of the problem of obtaining an effective molecule. In fact, automatic classification methods such as neural networks that identify "druglike" molecules are receiving much attention.

Promising Avenues

We cannot escape the question of whether protein bioinformatics is relevant to human health and whether it can effectively accelerate the drug discovery process. We can try to answer the question. A drug discovery process consists in several steps:

- Identification of the target
- Finding of suitable inhibitors
- Selection, among the potential inhibitors, of molecules with suitable druglike properties (lead compounds)
- Optimization of the lead compounds
- Laboratory testing of safety and efficacy
- Clinical trials

Phase 1 clinical trials involve healthy volunteers and attempt to determine whether the metabolic and safety parameters obtained in laboratory testing can be extrapolated to humans. Phase 2 trials involve a few hundred patients and attempt to establish the dosage regimen and provide a better understanding of the pharmacokinetic properties of the drug. During phase 3 trials, compounds are tested on a few thousand patients and their efficacy is evaluated and compared with that of similar products on the market, if any. Often, this phase includes blind testing. Patients are divided into test and control groups. The latter are administered placebos or previously available drugs for the same pathology. A positive evaluation of the results of Phase 3 trials allows the drug to be marketed. During phase 4 trials, the efficacy and safety of drugs are monitored after their introduction to the market.

The average time for the development of a drug ranges between 12 and 15 years and involves a cost of several hundred million dollars. The rate of expenditure increases during the process. A failure late in the process can have disastrous economical consequences. Late failure is not at all unlikely; only 1% of the drugs that enter clinical trials are estimated to end up on the market.

Where can bioinformatics intervene and accelerate or ameliorate the process? Certainly it can in target identification. Whether or not more targets are really needed is still debated in pharmaceutical companies. The overall number of targets for pharmacological intervention is very limited, and, in this case, some argue, more is not necessarily better. However, even if the biological processes that are being targeted by current drugs are sufficient and no need exists, at least from an economical point of view, to expand their repertoire, the question remains as to whether the specific molecules that are being targeted are the best ones. More "moonlighting" proteins are being discovered. This aspect of bioinformatics can be of invaluable help, especially if we can draw the complete, or semicomplete, scheme of the molecules involved in every biological process. In this case, we can select target molecules that are the most suitable to specifically interfere with the process, rather than randomly select them, as has been previously done.

The possibility of predicting the occurrence of undesirable effects that can manifest themselves later in the drug discovery process is of enormous interest and is being addressed by methods such as transcriptomics and proteomics, as well as by more recently developed automatic learning methodologies. Often, the discussion centers on whether or not any of the available drugs has been identified solely by computational methods, but posing the problem in such terms is not useful. Drugs must target biological processes, by and large mediated by proteins, and the understanding of the sequence–structure–function relationship in these molecules must form the basis of any modern discovery. Personalized medicine will no doubt be common in the future, and this future will only be made possible by the availability of a large set of data annotated in a detailed, accurate, and robust manner.

Suggested Reading

Ikubinyi, H. Drug research: myths, hype and reality, *Nature Rev. Drug Discov.* 2, 665–668, 2003.

Davis, A., Teague, S., and Kleywegt, G. Application and limitations of X-ray crystallographic data in structure-based ligand and drug design, *Angew. Chem. Int. Ed.* 42, 2718–2736, 2003.

Pastor, M., and Cruciani, G. A novel strategy for improving ligand selectivity in receptor-based drug design, *J. Med. Chem.* 38, 4637–4647, 1995.

Morris, G.M., Goodsell, D.S., Huey, R., and Olson, A.J. Distributed automated docking of flexible ligands to proteins: parallel applications of AutoDock 2.4, *J. Comput. Aided Mol. Des.* 10, 293–304, 1996.

Bohm, H.J. Prediction of binding constants of protein ligands: a fast method for the prioritization of hits obtained from de novo design or 3D database search programs, *J. Comput. Aided Mol. Des.* 12, 309–323, 1998.

Kramer, B., Rarey, M., and Lengauer, T. Evaluation of the FLEXX incremental construction algorithm for protein-ligand docking, *Proteins* 37, 228–241, 1999.

Stahl, M. and Rarey, M. Detailed analysis of scoring functions for virtual screening, *J. Med. Chem.* 44, 1035–1042, 2001.

Ewing, T.J., Makino, S., Skillman, A.G., and Kuntz, I.D. DOCK 4.0: search strategies for automated molecular docking of flexible molecule databases, *J. Comput. Aided Mol. Des.* 15, 411–428, 2001.

Abagyan, R. and Totrov, M. High-throughput docking for lead generation, *Curr. Opin. Chem. Biol.* 5, 375–382, 2001.

Lyne, P. D. Structure-based virtual screening: an overview, *Drug. Discov. Today* 7, 1047–1055, 2002.

Schneider, B. and Böhm, H. Virtual screening and fast automated docking methods, *Drug Discov. Today* 7, 64–70, 2002.

Braun, W. Distance geometry and related methods for protein structure determination from NMR data, *Q. Rev. Biophys.* 19, 115–157, 1987.

Allen, F.H. The Cambridge structural database: a quarter of a million crystal structures and rising, *Acta Crystallogr. B.* 58, 380–388, 2002.

Problem 9

Protein Design

Introduction to the Problem

In protein structure prediction, we infer the three-dimensional structure of a protein from its amino acid sequence. Chemical and molecular biology techniques can also be used to synthesize proteins that do not exist in nature. Therefore, we can "design" an amino acid sequence, produce the corresponding protein, and test its properties (Figure 69).

Why is this issue important? Several biotechnological applications require specific properties, and a natural protein that has all the desired characteristics might not exist. However, this problem has attracted much attention for a more fundamental reason, namely, the need to gain a better understanding of the sequence–structure relationship in proteins. The proteins we observe today are products of evolution, and, therefore, we can only observe those sequence–structure combinations that happened to arise during the development of life. In addition, several constraints act on the proteins we can observe in nature, because of their function, because of compartmentalization, and, in general, because of all those properties required for the survival of a cell. If we design an amino acid sequence and analyze its structure, we can derive rules genuinely related to the ability of a polypeptide to assume a given three-dimensional arrangement in space, not superimposed to any other constraint. Such knowledge can improve our ability to predict the structure of natural proteins.

Is designing a protein technically feasible? We have seen that evolutionarily related proteins preserve their structure during evolution and also that, in many cases, apparently unrelated protein sequences can give rise to similar structure. This observation implies that the structure–sequence code is degenerate: many sequences seem to be able to fit a given topology.

What is our chance of selecting a sequence that can fold into a given structure? For a protein of 100 amino acids, 20^{100} or 10^{130} possible amino acid sequences are possible. Approximately 1 out of 10^4 sequences is able to fold (i.e., has an energy landscape with a single global minimum). Therefore, we might have 10^{126} foldable sequences. From the statistical analysis of known protein structures, the number of different possible folds is estimated on the

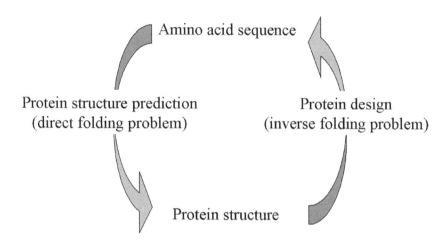

FIGURE 69
The direct and inverse folding problem.

order of 1,000; 10^{123} sequences should be able to adopt any given fold. Even if this figure were grossly overestimated, we should still be left with a large number of suitable sequences for each given fold. The conclusion can be drawn that the inverse folding problem (i.e., the problem of designing a sequence that fits into a given structure) should be easier to solve than the direct folding problem (i.e., inferring the structure of a given sequence).

This optimistic view generated a flurry of attempts, from as early as the beginning of the 1980s, to design novel proteins. These activities were directed towards both the redesign of the sequence of a known protein and the design of sequences able to adopt a novel, not yet observed, topology or fold. Some of the attempts in the former category met with considerable success, but realization came quickly that the problem was far from trivial. We have at least two hurdles in our path to protein design. One hurdle is the problem of precisely calculating the energy of a given protein conformation. For a successful design, we need a sequence that can stabilize the desired fold but none of the possible alternative folds. The other hurdle is the problem of evaluation of our results. Given the target experimental structure, we can relatively easily measure the correctness of a structure prediction, and, therefore, methods can be compared, combined, and improved in a stepwise approach. However, evaluating the extent of success of a protein design experiment is much harder. If the design is completely successful, the protein will be stable, soluble, and amenable to experimental structural analysis by X-ray crystallography or nuclear magnetic resonance. If the design is not perfect, the protein will not assume a nativelike structure and, even if the design is almost correct, we might not be able to characterize the result, learn about the mistakes, and improve the design.

Protein design must be based on what we know about protein structures, and the driving force of protein folding is mainly the entropic contribution of hydrophobic interactions. In fact, the core hydrophobic residues are the most conserved in protein evolution, and this condition highlights the relevance of their role in specifying the final structure. Not surprisingly, therefore, protein design attempts have paid special attention to the pattern of hydrophobic residues in the core of the model structure. Evidence from several experiments underscores the very important finding that at least the approximate fold a sequence will assume is dictated by its hydrophobicity pattern.

Other interactions, such as hydrogen bonds, play important roles. Potential donor and acceptor atoms form hydrogen bonds with the solvent in the unfolded structure. If left unpaired in the folded structure, the atoms would create energetically unfavorable conditions. Electrostatic interactions do not seem to substantially contribute to folding stability, as burying and, therefore, desolvation (removal of interactions with the polar solvent) of charged residues is destabilizing, but they can have a role in destabilizing the many alternative conformations accessible to a sequence with a given pattern of hydrophobic residues. The topology (i.e., the arrangements of the secondary structure elements) can depend on the sequence and propensity of the connecting elements, the loops whose prediction is a great challenge in protein structure prediction.

A simple binary pattern of hydrophobic and hydrophilic residues can be used to specify, for example, the conformation of an α-helix. In several successful designs of helix bundles, the sequence of each helix was created by the imposition of a periodicity of 3.6 on the hydrophobicity of its sequence. In this case, a stepwise approach can be used. Helices can be designed separately and their intramolecular interaction analyzed and joined into a single sequence in which helices form a two-helix, three-helix, or four-helix bundle. The specific topology depends upon the details of the interface between the helices and, therefore, on the specific amino acids rather than only on their hydrophobicity. The design of β-strand proteins is more complex. In this case, isolated β-strands are not stable (the hydrogen bond pattern of a β-sheet involves residues on different strands and is, therefore, nonlocal), and the stepwise approach is not feasible.

All these considerations should be taken into account in the protein-design process, both when it is based on biochemical intuition and when it is obtained via automatic algorithms.

Intuitive Design

The many attempts at designing novel proteins based on educated guesses and biochemical intuition are impossible to describe. Several research groups dedicated their efforts to the problem for several years. These often specialized in one of the many flavors of protein design, such as approaching the stepwise design of α-helical proteins, trying to obtain new topological arrangement of β-strands in β-sheet proteins, or reshuffling and rewriting the amino acid sequence of known proteins.

In 1986, Chris Sander organized a course on protein design, a rather novel idea at the time. The participants, including several renowned scientists, were divided in groups. Each group worked on a different design project. The experiment was repeated twice, in 1990 and in 1994. Only rarely was the ability to fold the designed proteins tested experimentally. Nevertheless, a mention of which projects were selected in these occasions is instructive because they are representative of what was being tried in this area.

Several attempts were made to obtain idealized folds (i.e., optimized architectures without the asymmetries and local distortions present in natural protein structures), as well as to create designs that would "minimize" the size of known proteins to verify the extent to which the peripheral parts of the structure play a role in function and stability. Many protein structures contain repetitions of basic folding units, and several attempts were made to design them by using blocks with the same sequence.

The most popular architectures used as starting points for protein-design projects were the antiparallel four-helix bundle and the so-called TIM (triosophosphate isomerase) barrel. An α-helical bundle is an arrangement of two, three, four, or more helices that can be parallel or antiparallel. All these topologies are observed in natural proteins, and protein design has been especially useful in helping us understand their sequence structure relationship in detail.

The prototype of the α-helical bundle is the protein Rop (repressor of primer). This protein is formed by two identical chains, each formed by two antiparallel α-helices that spontaneously assemble to form the bundle. The self-assembly of the two chains and the rather simple architecture of each of them prompted several attempts to redesign the protein as a single chain, as a bundle of identical helices, as a protein with different connectivity between the helices, and with loops of different length. As mentioned before, most of these attempts took advantage of the possibility of designing the proteins in a modular way, starting from helices, then continuing to two chains of two helices and finally to the full four-helix bundle. The results of these experiments and others like them led to a better understanding of the geometry of amino acid helices and highlighted some interesting aspects of the sequence–structure relationship. The examples described below illustrate how useful these experiments have been in revealing that the quaternary structure of helical proteins can be modulated by small sequence variations, which are achieved in nature not only by the stabilization of the "desired" topology but also by the destabilization of the alternative ones.

First, we introduce the "helical wheel" representation of a helix. The wheel is a schematic projection of the Cαs on a plane perpendicular to the helix axis. The residues will be indicated with the letters *a* to *g*. The relative position of two parallel and antiparallel helices is shown in Figure 70.

The amino acids in positions *a* and *d* should be hydrophobic, because they are buried within the structure. However, the pattern of their interaction in

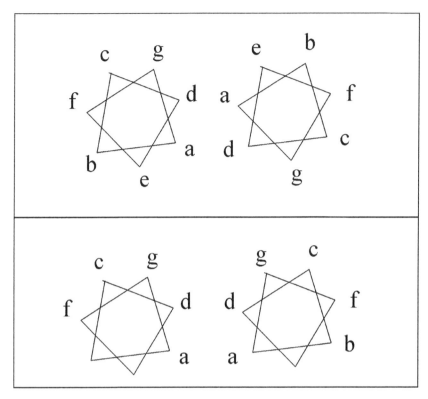

FIGURE 70
Two parallel (top) and two antiparallel (bottom) helices packing against each other.

the two cases is different, and this difference can be used to differentiate between the two topologies. An intriguing problem arises when the expected patterns of interaction between parallel two- and three-helix bundles are compared. Also, in this case (Figure 71), the amino acids in positions *a* and *d* should be hydrophobic. How does nature specify the correct arrangement? A combination of protein-design experiments and analysis of known structures showed that a solution to the problem can be achieved by destabilizing one of the two architectures. If a polar residue is in position *a*, its presence in the interior of the protein destabilizes both arrangements. However, in the three-helix bundle, position *a* forms more interactions with hydrophobic residues than in the two-helix bundle, so it is comparatively more destabilized by this "imperfection" than the paired helix architecture. In other words, nature trades stability for specificity.

Protein-design experiments also showed that, in parallel helix bundles, if position *a* is an isoleucine and position *d* is a leucine, the helices tend to form dimers, whereas in the reciprocal situation (position *a* is a leucine and position *d* is an isoleucine), the tetrameric assembly is favored. This apparently puzzling observation can be explained by the simplified representation of the helices (Figure 72). In a dimer, position *d* of one helix interacts with

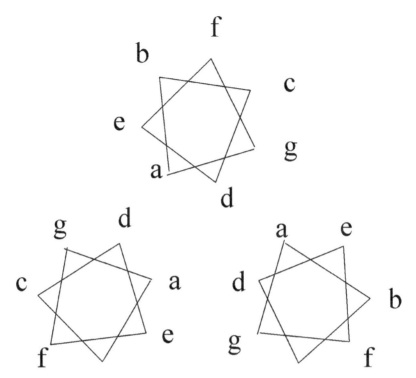

FIGURE 71
A three-helix bundle.

position d of the other helix, and the two amino acids are rather close in space. The presence of two bulky isoleucines disfavors such an arrangement, whereas leucine residues can be accommodated more easily. Once again, a specified quaternary structure interaction can be favored by destabilizing alternative arrangements. These results, and many more described in scientific reports, illustrate how protein design can highlight properties of amino acid sequences related to fundamental properties of proteins.

Another popular protein architecture repeatedly used in protein-design experiments is the TIM barrel (Figure 73) (see color insert after page 40). It is formed by a central barrel of eight β-strands surrounded by eight external α-helices. Proteins with this fold usually show some deviation from the basic architecture. For example, they can contain excursions of the chain to form other secondary structure elements, the helices and the strands can have different length, or the connecting loops can be different. However, they are all formed by a repetition of the basic βαβ folding unit, formed by two parallel β-strands connected by an α-helix packing against the strands, a very tempting modular arrangement for protein design. Indeed, idealized TIM barrels, with the same sequence in each of the units, have been designed, as have "reduced TIM barrels," formed by only four strands and four helices. Circular permutation of the unit sequences (moving the sequence of the first unit at

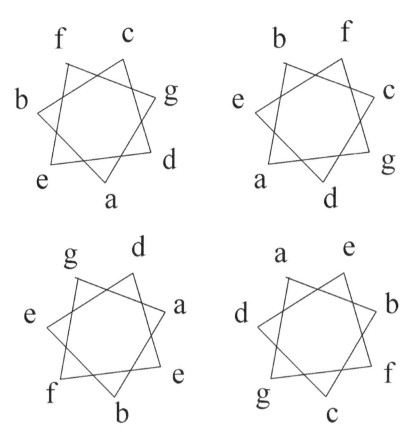

FIGURE 72
A four-helix antiparallel bundle.

the end or the sequence of the last unit at the beginning), as well as attempts to add extra units at the beginning or the end of the protein, have been tried.

This flurry of design projects slowed after a few years. Possibly, scientists were becoming frustrated by the difficulties of the problem. Although the design of helix bundles and, in some cases, the successful redesign of the sequence of existing folds were possible, the challenging result of a completely novel fold entirely designed *de novo* seemed to be still far beyond our reach.

Lattice Models and Automatic Methods

An intrinsic limitation of intuitive design is that the reasons an experiment is successful might be difficult to pin down, and, therefore, intuitive design does not necessarily teach us sufficiently general rules. We might understand

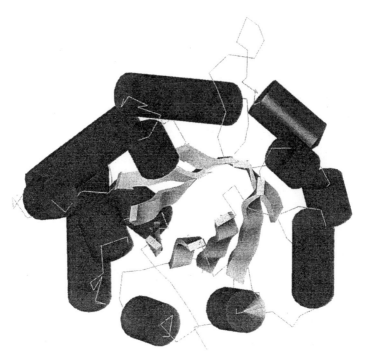

FIGURE 73
A TIM barrel (PDB id: 8TIM).

some underlying principle, as in the case of helix bundles, but not always. Conversely, if we could devise an automatic method, its effectiveness could be tested and evaluated and this method could be more effective in teaching us the underlying rules governing protein folding.

Some fundamental aspects of protein folding and design have indeed been captured by automatic, extremely simplified, and apparently completely unrealistic models: the lattice models. The idea is to analyze sequence–structure relationships in a simplified and ideal case in which the sequence is made of very few "symbols" endowed with some property, and the "structure" is a path in a two-dimensional or three-dimensional grid. The aim is to optimize some simplified "energy" function that describes the interaction between the symbols. The purpose of the approach is not to obtain a designed protein, but rather to answer more general questions: Which properties should the sequence of symbols have to be "designable," that is, have a single energy minimum? If we have an energy function, and we generate a set of arrangements for which the energy is known to be minimum, under which conditions can the arrangements be used to recover the input energy function? This question relates to the problem of how useful is the set of known protein structures for extracting the parameters of a force field. Some of the results obtained in this area are quite interesting.

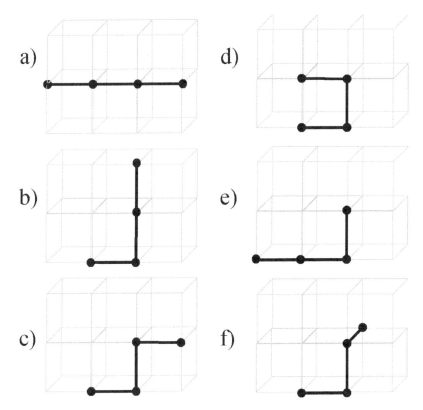

FIGURE 74
Six self-avoiding paths in a three-dimensional lattice for four elements.

The general idea is shown by the following simplified example: We have two types of "amino acids," one polar (P) and the other hydrophobic (H), and an energy function such that two Hs interact favorably if they are in adjacent positions in the grid but nonconsecutive in the sequence, whereas the H-P and P-P interaction is neutral. Is there a sequence of length four such that it can assume at least one structure in the grid that has an energy lower than any of the other structures?

If the dimension of the grid is small, we can enumerate all the possible "structures" in the grid, calculate their energy, and derive some conclusions. If we consider only non-symmetry-related, self-avoiding paths allowed for four residues, we obtain the six possibilities in Figure 74.

We can use sixteen possible combinations of P and H symbols, but six (shown in italics in Table 9.1) can be eliminated because, in this simplified case, we do not distinguish between the two possible directions of the chain. The conformation designated with (d) is the only one with a contact (between the first and last position), so we can compute the energy of this conformation for all possible sequences, assuming that a favorable contact (H-H) has an energy of −1 and all other pairs have an energy of 0. The conclusion here is

TABLE 9.1

All Possible Sequences of Hs and Ps for a Four-Element String and Their Energy When Folded As in (d) of Figure 74.

Position 1	Position 2	Position 3	Position 4	Energy in Conformation e
p	p	p	p	0
p	p	p	h	0
p	p	h	p	0
p	p	h	h	0
p	h	p	h	0
p	h	h	p	0
p	h	h	h	0
h	p	p	h	−1
h	p	h	h	−1
h	h	h	h	−1
p	h	p	p	0
h	p	p	p	0
h	p	h	p	0
h	h	p	p	0
h	h	p	h	−1
h	h	h	p	0

that the structure indicated with (d) is "designable" and that more than one sequence can be used to obtain it.

This simple example shows that, given an energy function, if we can enumerate all the possible patterns for a sequence of a certain length, we can compute which sequence path has the optimal energy and also whether the combination is unique or, as in our example, degenerate in the sense that more sequences can fit the same structure.

The number of possible paths and sequences increases with the length of the sequence (Table 9.2), and the brute-force approach depicted here cannot be used. Three-dimensional grids have been, therefore, coupled with optimization algorithms, such as simulated annealing or Monte Carlo, and have been extensively used to derive patterns of Hs and Ps able to specify, for example, structures similar to α-helices or to bundles of α-helices. The resulting hydrophobicity pattern can be compared with the hydrophobic and hydrophilic pattern observed in real protein structures, and, in general and for most of the methods, the computed and observed patterns do not differ substantially. Lattice models with increasing complexity have been implemented and even extended to real Cα protein coordinates.

Lattice models can also be used to answer the inverse question. For example we can select a "sequence" (i.e., a string of Hs and Ps), try all possible conformations in a lattice, and rank their energy. This process can tell us whether the given sequence stabilizes more than one structure and, therefore, is not suitable for a design experiment (because its energy landscape contains more than one energy minimum). Calculations of this type have been used, for example, to estimate the number of foldable sequences and also to conclude that some "structures" represent the minimum energy for many

TABLE 9.2

H and P Element Sequences and Possible Conformations in a
Three-Dimensional Grid for Different Sequence Lengths

Number of Elements	Number of Sequences (2^N)	Number of Possible Paths in a 3D Grid
4	16	10
5	32	20
6	64	36
7	128	72
8	256	136
9	512	272
10	1024	528
50	$>10^{15}$	$>10^{14}$

sequences, whereas others are stabilized by a small number of sequences. The sequence should contain as many Ps as possible, with Hs in points of the grid that make multiple interactions and without internal ("buried") Ps. These conclusions sound very sensible and not very dissimilar from what is observed for real protein structures, although the limitation of their very low level of detail is obvious.

Nature uses subtle differences between amino acids to specify the final native structure. Therefore, these methods can only provide a first-level solution to the problem and highlight trends rather than properties, but they have proved useful in helping us understand some basic rules of the overall architectural space available to protein structures and, together with the results of intuitive design projects, in highlighting some of the properties that any automatic method for protein design should take into account. In general, the following conclusions should be considered:

- The hydrophobic pattern of the designed sequence
- The solvent accessibility preference for each amino acid
- The amino acid preferences for secondary structure elements
- The specific interactions able to generate asymmetry and destabilize alternative folds

The final goal is to optimize an energy function. Here, as in protein structure prediction, both molecular mechanics and pairwise potential can be used, together with optimization techniques such as Monte Carlo, genetic algorithm, and dead-end elimination.

The most difficult aspects to be taken into account, here as in intuitive design, are the destabilization of alternate folds and the capability of the designed sequence not only to stabilize the desired fold but also to have a folding pathway that leads to it. The second aspect is beyond our present capabilities because of our limited understanding of the actual process of folding. As far as the first aspect is concerned, we can take advantage of fold recognition methods to verify whether the final selected sequence is likely to fit other folds present in our protein structure database.

The cross-fertilization between protein structure prediction and protein design is not limited to the use of fold-recognition methods at the last stage of the procedure. For example, methods for the prediction of protein structures that do not share any sequence or structural similarity with known proteins have been successfully applied to protein design by cycles of the following protocol:

- Selection of an initial topology for the target structure
- Definition of a set of distance constraints that define the desired fold
- Generation of several initial conformations from fragments of proteins of known structure on the basis of the distance constraints
- Sequence optimization for each of the starting structures by application of Monte Carlo simulations
- Prediction of the structure of the selected sequence by use of fragment-based methods

Baker et al. used this strategy to produce a soluble, monomeric, stable protein with a novel fold, the X-ray structure of which showed an impressive similarity with the designed model. Interestingly, in this successful case, no attempt was made to destabilize alternative putative folds, and the kinetics of folding were not taken into account in the design. Also, the similarity between the designed and experimentally determined structure is much higher than the accuracy achieved by the same method when it is used to predict protein structures. One possible reason for this difference is that the designed protein does not have any functional constraints and, therefore, it is more "regular" than a real natural protein, in which the need for a functional site can make the relationship between sequence and structure more complex. Protein active sites are most likely suboptimal in terms of their structure because they need to satisfy functional and dynamic constraints. Therefore, predicting their conformation on the basis of energy optimization can be more difficult. However, in several cases, active sites have been successfully inserted into known proteins, as we will see in the next problem, and nothing, at least in principle, prevents the same from being done in a designed protein. Such an accomplishment could lead to the engineering of tailored molecular machines.

Promising Avenues

The fields of protein structure prediction and protein design have crossed paths at different times, with beneficial effect. Pairwise potentials, as well as methods to evaluate sequence–structure fitness, initially developed for structure prediction, have found useful applications in protein design. More recently, fragment-based methodologies have been shown to be equally "portable" from one field to the other. This cross-fertilization can be expected to continue, and protein design will profit by innovative ideas in protein structure prediction. At the same time, successful protein-design experiments will be of invaluable help in understanding the nature of the relationship between protein sequence and structure.

The understanding of protein evolution is one of the most powerful tools for assigning structure and function to unknown proteins, but the problem of distinguishing regions that are conserved because of functional constraints from those required for function persists. In design, the problems of structural stability and function can be separated, which can highlight important features of protein sequences. Therefore, in the future, one can expect that the boundaries between the two fields will become increasingly less clear, at least from the point of view of the methods employed.

We now have a handle on the design of novel stable proteins on one hand and on the redesign of functional sites on the other hand (see next problem). Combination of the two aspects will allow us to specify our requirements and design and produce nanomachines that perform predetermined functions.

Probably more in this field than in most of the areas we have described so far, a collaborative effort between theory and experiment is needed, as is evidenced by the fact that the successful cases of protein design have all arisen from a close interplay between computational, structural, and molecular biology.

Another aspect that is already receiving attention but should be further pursued is the large-scale design of molecular partners. Can we exploit what we understand about molecular interactions to design molecules that interact with a target protein? Such a technology would be instrumental in many fields and have an impact on both human health and biotechnology.

Suggested Reading

Crippen, G.M. Prediction of protein folding from amino acid sequence over discrete conformation spaces, *Biochemistry* 30, 4232–4237, 1991.

Jones, D.T. De novo protein design using pairwise potentials and a genetic algorithm, *Protein Sci.* 3, 567–574, 1994.

Desjarlais, J.R. and Clarke, N.D. Computer search algorithms in protein modification and design, *Curr. Opin. Struct. Biol.* 8, 471–475, 1998.

Mirny, L., and Shakhnovich, E. Protein folding theory: From lattice to all-atom models, *Annu. Rev. Biophys. Biomol. Struct.* 30, 361–396, 2001.

Sander, C., Vriend, C., Bazan, F., Horovitz, A. Nakamura, H., Ribas, L., Finkelstein,
    A.V., Lockhart, A., Merkl, R., Perry, J.L., Emery, S.C., Gaboriaud, C., Marks, C.,
    Moult, J., Verlinde, J.C., Eberhard, M., Elofsson, A., Hubbard, T.J.P., Regan, L.,
    Banks, J., Jappelli, R., Lesk, A.M., and Tramontano, A. Protein design on com-
    puters. Five new proteins: Shpilka, grendel, fingerclasp, leather, and aida, *Pro-
    teins* 12, 105–110, 1992.
Pessi, A., Bianchi, E., Crameri, A., Venturini, S., Tramontano, A., and Sollazzo, M.
    A designed metal-binding protein with a novel fold, *Nature* 362, 367–369, 1993.
Hill, R.B., Raleigh, D.P., Lombardi, A., and DeGrado, W.F. De novo design of helical
    bundles as models for understanding protein folding and function, *Acc. Chem.
    Res.* 33, 745–754, 2000.
Main, E.R., Jackson, S.E., Regan, L. The folding and design of repeat proteins: Reach-
    ing a consensus, *Curr. Opin. Struct. Biol.* 13, 482–489, 2003.
Kuhlman, B., Dantas, G., Ireton, G.C., Varani, G., Stoddard, B.L., and Baker, D. Design
    of a novel globular protein fold with atomic-level accuracy, *Science* 302,
    1364–1368, 2003.
Tramontano, A. A brighter future for protein design, *Angew. Chem. Int. Ed. Engl.* 4325,
    3222–3223, 2004.

Problem 10

Protein Engineering

Introduction to the Problem

Can we modify a native protein sequence and endow it with novel properties? The solution to this problem requires a combination of the tools we described for both protein-structure prediction and protein design. We must first design a modified protein sequence that is able to carry out the desired function, and then we must predict the structural effect of the mutations we introduced in the protein.

Similar protein sequences have similar structures, but this knowledge is not very helpful in protein engineering, because it only applies to naturally evolved proteins; that is, proteins containing mutations that have been accepted by evolution because they are not deleterious and, therefore, not expected to substantially affect the protein structure. On the contrary, if we artificially modify even a single amino acid of a protein sequence, we cannot be sure that the structure will be preserved.

In general, we should solve the folding problem for the new sequence and compute the conformation that corresponds to its global free-energy minimum, without even being certain that such a conformation exists. In other words, predicting the structure of a mutant can be even more difficult than predicting the structure of a native protein. Furthermore, if we want to endow the protein with a specific biochemical function, we must be aware that function is often mediated by clusters of residues that must be in a very precise relative location to be functional. Therefore, our prediction of the details of the protein structure must be very accurate.

One more hurdle in the process is that functional residues are not necessarily optimal for structural stability. In a number of cases, mutations of active site residues reduce activity but increase stability of the protein. This finding suggests that the requirement of having specific residues in active sites might introduce "energetic defects" into a protein. Therefore, our energetic calculations can be completely inappropriate for this task.

Nevertheless, engineering novel properties in natural proteins has a number of important applications, such as testing our understanding of the functional site properties, creating diagnostic tools, and producing biotech-nologically useful proteins.

Combining Functions

Existing functions can be combined by fusing two or more proteins or protein domains together to obtain a new function. An application of this approach is the yeast two-hybrid method that we used to fuse our proteins to the binding and activation domains of the Gal4 gene product.

In general, combining functions requires the design of a linker sequence that joins the domains without altering their structure and function. If we are dealing with one or more proteins of known structure, an inspection of the three-dimensional arrangements can help us decide the length of the linker and whether to insert the extra domain at the amino or the carboxy terminus of the protein. We want the linker to be flexible and hydrophilic. The most flexible amino acid is glycine. It does not have a side chain and, therefore, has fewer limitations on the values its ϕ and ψ angles can assume. Surveys of long connecting regions between domains of known proteins suggest that sequences of glycines and serines (the smallest hydrophilic amino acids) are ideally suited for this task.

This procedure is commonly applied to antibodies. Only one light-chain domain and one heavy-chain domain are involved in antigen binding, and these domains are conveniently fused into a single chain to eliminate concern about their quaternary assembly to obtain the complete binding site. This fusion is achieved by cloning the heavy-chain variable domain, followed by a long stretch of serine and glycine amino acids, followed by the light-chain variable domain.

Global Properties

Often, proteins of biotechnological or pharmacological interest are not opti-mal. They might be too unstable or insoluble, or they can be too labile at higher temperature. For example, an enzymatic activity is often used in powdered laundry detergents. However, we do most of our laundry at a temperature higher than room temperature, at which most enzymes do not function efficiently or do not function at all. The problem can sometimes be solved by addition of a protein with the desired stability at high temperature. The protein may be found in thermophilic organisms such as bacteria that live in volcanic areas. When a natural protein with the desired stability cannot be found, the sequence of an existing mesophilic (i.e., nonthermophilic) enzyme can be modified to achieve higher thermal stability. The obvious procedure is to analyze families of proteins that contain both thermophilic and mesophilic members and derive rules that distinguish the former from the latter. This approach is less successful than we would expect. Some rules have been derived, but the specific sequence features that confer thermal stability to a protein are extremely difficult to identify. This difficult is probably caused by the fact that thermal stability has evolved independently in different protein families, and, therefore, different proteins rely on different evolutionary strategies.

However, stability is usually correlated with thermophilicity. In other words, proteins in which the difference in energy between the unfolded and folded state is higher tend to be more thermally stable. Therefore, we can attempt to increase the thermodynamic stability of the protein under examination. One way to do so stems from the concept of pairwise potentials: sequence patterns that are observed more often are likely to be more stable. Thus, if we have a family of protein sequences, the most observed amino acid in each position can be assumed to be the most stabilizing one, and the proteins that contain the most observed amino acids are expected to be the most stable proteins. This strategy relies on the hypothesis that the effect of each mutation is independent (i.e., that amino acids do not contribute to stability in a cooperative manner) and on the availability of a good multiple alignment of the protein family.

Another possibility is to highlight less than optimal interactions in the protein structure, either by simulating every possible substitution in a set of candidate positions and evaluating the energy of the resulting structure or, as in "intuitive design," by application of biochemical intuition.

Increasing solubility in aqueous solvent is another important goal of many protein-engineering projects. Once again, an examination of the variability in the sequences of a family or simulated replacements of hydrophobic amino acids with hydrophilic amino acids and computing the energy can identify positions that can accept more hydrophilic substitutions.

Another way to increase the stability of a protein is to introduce covalent bonds between different parts of the molecule. The side chains of the amino acid cysteine can form covalent bonds with each other, as shown in Figure 75 (see color insert after page 40), and this bonding can increase the stability of a protein. A protein structure database analysis can be used to evaluate the optimal values of the angles and distances between the cysteine atoms in experimentally observed disulfide bridges. Subsequently, the protein structure can be scanned for positions that, once replaced with cysteines, could form disulfide bridges.

The favorable positions must be accurately selected because the stability that we gain depends upon the difference in the energetic contribution of the disulfide bridge in the folded versus the unfolded state. Therefore, we increase the stability of the engineered protein only if the geometrical constraints provided by the three-dimensional structure allow an arrangement

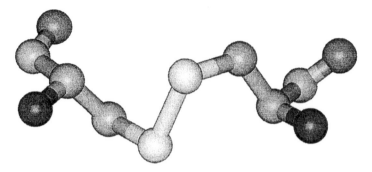

FIGURE 75
A disulfide bridge.

of atoms involved in the bridge that is more energetically favorable than in the unstructured state.

Some methods take into account the local environment of each of the cysteines. They are usually based on neural networks trained on a data set of sequences that contain cysteines known to form disulfide bridges.

Active and Binding Sites

The most ambitious goal of protein engineering is the design of novel catalytic or binding functions in proteins of known structure. The difficulty arises from the fact that, in general, catalysis is brought about by the cooperation of a few amino acid side chains, exactly positioned to perform their function. Not much room is allowed for mistakes.

Many important biochemical reactions involve metal atoms such as zinc, calcium, iron, and magnesium, and several structural motifs can bind these elements. A first step in designing an enzyme-active site is to learn how to modify an existing protein to create a metal-binding site in its structure. This type of experiment has met with success in several cases. The obvious method is to locate in the target protein residues whose backbones can be superimposed with residues that coordinate a metal in a known metal-binding site. The electrostatic properties of protein atoms that do not directly coordinate the metal, but surround the binding site, can play a role, and some methods take them into account. Automatic learning methods can be applied to this process by the use of interresidue distances of atoms that surround the experimentally determined metal-binding site to train the system.

Perhaps the most studied case in the engineering of active sites is that of serine protease. The triad of amino acids that perform the catalysis can be regarded, on first approximation, as an independent motif. We find these amino acids in proteins with architectures as diverse as those of serine

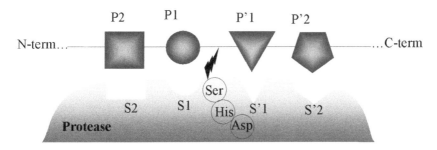

FIGURE 76
The nomenclature for serine proteases. The amino acids of the substrate are indicated by S, and the corresponding recognition sites on the surface of the enzyme are indicated by P.

proteases and lipases. Furthermore, we know the structure of several enzymes of the family, and this knowledge has led to a deep understanding of the sequence–structure–function relationship. Not only the geometry of the catalytic triad is well known, but the surrounding regions, which bring about the specificity of the enzymes, have also been thoroughly analyzed.

The nomenclature used in this field is shown in Figure 76. Both the residues and the sites are numbered starting from the cleavable bond and working outward. Primed letters refer to the carboxy-terminal part of the substrate and to the corresponding sites. The S1 residue is the major determinant of specificity, and it is recognized by the P1 site (or pocket) of the enzyme. A change of one or more of the residues lining the P1 pocket can change the specificity of the enzyme. This strategy has been successful in many cases.

In the case of the hepatitis C virus protease that we have described before, the structure of the enzyme was predicted by comparative modeling on the basis of a very weak sequence similarity with other proteases and a not necessarily reliable sequence alignment. Protein engineering was used to test the model by verifying whether it was able to predict the effect of specific mutations in the protein (Figure 77).

The authors designed a two–amino acid mutant that, if the model was correct, would switch the specificity of the protease from P1 = cysteine to P1 = phenylalanine. The success of the experiment provided a validation of the model, at least in the region of the active site. This work shows that successful protein-engineering experiments can also be designed on the basis of a comparative model (i.e., even in the absence of an experimentally determined structure).

Catalytic Antibodies

Another more general and very interesting strategy has been applied to the design of enzymatic activities. To understand the method, we must recall

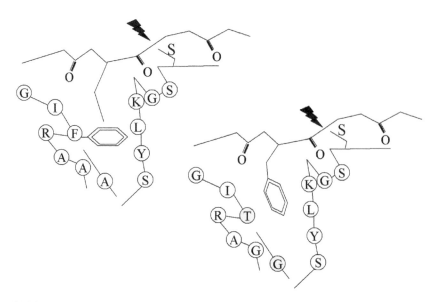

FIGURE 77
The engineering of the hepatitis C protease specificity. The model (left) suggested that the presence of a bulky phenylalanine was responsible for a small S1 pocket. The enzyme indeed recognizes a small amino acid (cysteine). By replacing the phenylalanine and an alanine just below it with a threonine and a glycine, the enzyme acquired the ability to accommodate, and, therefore, recognize, a phenylalanine in its P1 pocket.

that an enzyme catalyses a reaction by lowering the energy of the transition state, which is a high-energy intermediate of the reaction. This concept can be understood by considering the very simple reaction

$$H + H_2 \Leftrightarrow H_2 + H$$

Let us rewrite the reaction as

$$H_A + H_B\!-\!H_C \Leftrightarrow H_A\!-\!H_B + H_C$$

In this reaction, the reagents (H_A and $H_B\!-\!H_C$) have the same energy of the products ($H_A\!-\!H_B$ and H_C). We can now visualize the course of the reaction if we imagine the proton and the hydrogen molecule coming progressively closer to each other. The energy of the system increases as the molecules approach each other and becomes maximal when the H_B proton is equidistant from H_A and H_C. The course of the reaction is depicted in Figure 78.

In general, in a biochemical reaction, reagents and products can have different energy, but one high-energy intermediate always occurs along the course of a reaction. The job of an enzyme is to lower the energy of this high-energy intermediate and, thereby, accelerate the rate of the reaction, as can be seen in Figure 79. The enzyme lowers the transition barrier by binding

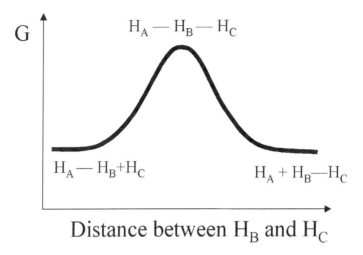

G

$H_A \!-\! H_B \!-\! H_C$

$H_A \!-\! H_B \!+\! H_C$

$H_A \!+\! H_B \!-\! H_C$

Distance between H_B and H_C

FIGURE 78
The reaction $H_A + H_B\!-\!H_C \Leftrightarrow H_A\!-\!H_B + H_C$.

and stabilizing the high-energy intermediate. One very imaginative way to achieve this effect is to raise an antibody against a chemical structure that mimics the transition state. One of the first attempts to do so had as its target the hydrolyzing reaction of serine proteases which was achieved by injecting a mouse with a molecule resembling the reaction transition state. The antibodies extracted from the mouse were able to catalyze the hydrolysis reaction.

Many abzymes (antibodies endowed with catalytic activity) have been generated for many biochemical reactions by use of more sophisticated techniques than mouse immunization to obtain the desired molecule and, especially, by subsequently optimizing the antibody via rational redesign. The resolution of an increasing number of three-dimensional structures of abzymes has also brought a better understanding of the appearance and evolution of catalytic functions, not only in antibody binding-sites, but also in enzyme active-sites.

A further step in this direction is the so-called anti-idiotypic approach. Suppose we have an antibody A_1 against a protein antigen, and we generate a set of antibodies A_2 against this antibody. Then the assumption is that some of the A_2 antibodies will resemble the original antigen. In other words, the antibody A_1 will represent a negative image of the antigen, and some of the A_2 antibodies will have a binding site similar to the surface of the original antigen.

This idea can be exploited to generate abzymes. The A_1 antibody is raised against the active site of an enzyme, so that the structure of the binding site of some of the A_2 antibodies might resemble the active site of the original antigen and be able to carry on the same enzymatic reaction. This strategy has been successful in a few cases, and it has also been applied to the design

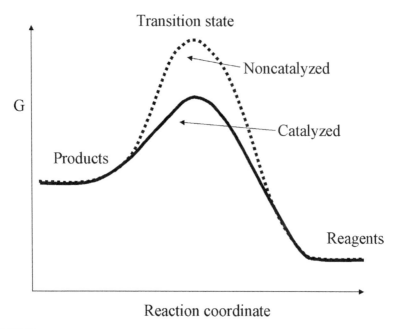

FIGURE 79
A catalyzed and a noncatalyzed biochemical reaction.

of metal-binding sites. The range of applications of antibodies has been made wider by the use of combinatorial systems that generate antibodies in vitro.

Combinatorial Design

The availability of designed proteins able to carry out a given function is important, not only because of their technological applications, but also because their analysis allows us to improve our understanding of the basis of enzymatic activity. This reasoning implies that we are not only interested in rational redesign of proteins, but also in observing the various ways in which a protein sequence can specify a given function. This goal can be achieved rather effectively by the use of combinatorial strategies. We describe only one of the many experimental approaches that have been devised, phage libraries, and as usual, we trade precision for simplicity.

Filamentous phages are viruses that infect a bacterial host cell. They exploit the cellular machinery to replicate their DNA and synthesize their proteins, after which they reassemble and exit the host. They are composed of a DNA genome surrounded by several proteins, as shown in Figure 80 (see color insert after page 40). The important features of the system that make it ideal for combinatorial display is the fact that foreign sequences can be cloned at

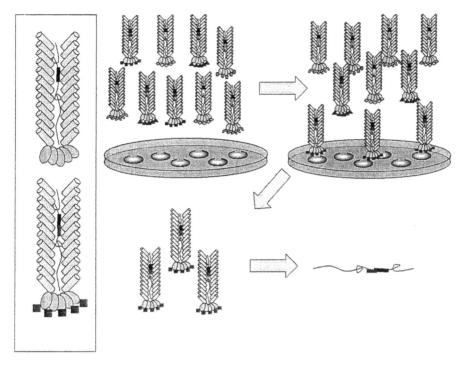

FIGURE 80

Phage display experiments. An oligonucleotide cloned upstream of the gene for the pV protein of the bacteriophage (shown as a gray oval) is exposed on the surface of the phage. If the cloned sequence is random, each phage will display a different peptide sequence. The phage population can be made to interact with a target protein. Only phage that contain a peptide that binds to the target protein will be retained, whereas the others can be washed away. The DNA sequence of the selected phage will directly reveal the sequence of the interacting peptide. The pVIII protein, shown as a gray cylinder, is also suitable for displaying foreign peptides on the surface.

the end of two of their genes, and the respective translated amino acid sequences are exposed on the surface of the complete phage particle.

The cloned segment can have a predefined sequence, but it can also be a randomly assembled subsequence of nucleotides. In this case, each phage colony displays a different fused peptide on its surface. A molecule of interest can be immobilized on a matrix or on a plate and made to interact with a population of phage. Only phage that display a sequence that binds to the selector molecule will be retained. These viruses can be collected, amplified by reinfecting host cells, and, finally, their DNA sequence can be obtained (Figure 80). The translation of the DNA sequence of the selected phage provides the sequence of the peptide sequence that binds to the selector molecule. Clearly, any protein-design experiment to introduce a binding property can take advantage of this methodology.

This technique is extremely powerful. For example, by using an antibody as selector, we can identify which oligopeptides it binds to, and these oligopeptides can be compared with the database of known sequences to identify the natural cognate antigen of the antibody. In principle, the method can be used to detect oligopeptides that bind any protein. In some cases, this technique can be used to find which of the known proteins, if any, binds to the target molecule. Unfortunately, rarely are interaction surfaces between proteins composed by linear segments. More often, a protein recognizes residues that are close to each other in the protein structure but not necessarily in the protein sequence. If the structure of the natural binding protein is known, an inspection of its surface can highlight regions that can be mimicked by the selected linear sequences. This inspection can help identify the location of the binding site.

An even more powerful application of phage libraries is the generation of antibodies. Single-chain antibodies, described previously, can be cloned at the end of the appropriate phage proteins, but, more importantly, a variety of different antibodies, each with different randomly generated sequences for their antigen-binding regions, can be cloned and selected for their ability to bind a given protein molecule. This process is a rapid way to generate antibodies that bind any given protein, and it has a wide range of diagnostic and biotechnology applications. The knowledge of the sequence–structure relationship in antibody-binding sites has been instrumental in the design of more effective libraries. The rational design of libraries to increase the likelihood of obtaining better binding molecules is a field of analysis that bridges protein bioinformatics (to select the target molecules and the sites to be randomized) with nucleic acids analysis (to design the best combination of nucleotides that gives the highest probability to obtain the desired set of sequences).

Dissecting the Folding Pathway of Proteins

The introduction of mutations into a native protein can also be used to improve our understanding of the folding process. We can envisage the folding of a protein as a reaction in which the reagent is the unfolded state and the product the native state. For many small proteins, the process is a simple two-state folding, with a single high-energy barrier that represents the transition state of the process. The speed at which a protein folds depends upon the height of the energy barrier, which is analogous to what occurs in a biochemical reaction.

Let us assume that we can introduce a mutation in our protein without perturbing its structure and without affecting the energy of the unfolded state. The mutation affects the stability of our protein, and we can measure the extent of the stability change. However, if the mutated residue is involved in early interactions in the folding process (i.e., if it is involved in interactions in the transition state) its mutation also affects the height of the barrier and, therefore, the folding rate of the process.

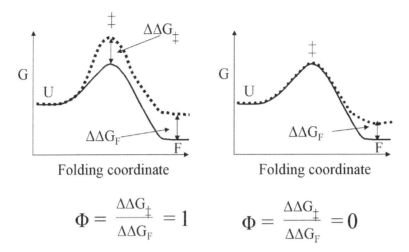

$$\Phi = \frac{\Delta\Delta G_{\ddagger}}{\Delta\Delta G_F} = 1 \qquad \Phi = \frac{\Delta\Delta G_{\ddagger}}{\Delta\Delta G_F} = 0$$

FIGURE 81
The definition of the φ value.

We can experimentally measure the change in the free-energy difference of the transition state (related to the folding rate) and the free-energy difference of the folded state (related to the stability of the protein). As shown in Figure 81, if the free-energy differences are similar, the mutated residue is involved in the transition state. The ratio between these two values is called the φ value for the mutated residue. A φ value of approximately 1 indicates that the residue is involved in interactions established early in the folding process, whereas a φ value of approximately 0 indicates that the amino acid only establishes interactions late in the folding process.

If we generate several single-residue mutants of a protein and compute their φ values, we can obtain a map of the residues that are involved in stabilizing interactions early in the folding process and try to understand this mechanism. Once a sufficiently large set of data has been collected, a number of questions can be posed: Is the transition state conserved through evolution? Do nonnative interactions occur in the transition state; that is, do interactions form early in the folding process that are not present in the final structure? Can we optimize the folding rate of a given protein? How do designed proteins fold?

Promising Avenues

Only in the past two decades have we been able to rationally modify proteins (i.e., to specifically mutagenize selected amino acids), but the technology has already pervaded our lives. The impact has been enormous. We can not only observe nature, but we can also perform experiments to test its mechanisms.

The role of specific amino acids or regions of proteins could be assessed by rationally modifying them and verifying the results. Historically, the first reported protein modification experiment involved a cysteine amino acid in the active site of a protein called tyrosil transfer RNA synthase, an enzyme that constructs the adaptor molecule that recognizes a base triplet on the RNA and adds the appropriate amino acid to the nascent protein chain. The experiment proved the specific role of the mutated amino acid in catalysis. This experiment was but the beginning. Most of what is written in biochemistry books about the role of specific amino acids in catalysis, for example, directly derives from mutagenesis experiments. Furthermore, the role of specific interactions in stability and folding has been tested with the same technique.

Things are not always so easy, however. Local interactions are not solely responsible for catalysis, and in many cases, the results were difficult to interpret, and very detailed studies about the effect of the cooperation of several sites were required.

The possibility of combinatorial selection for desired properties was another milestone in the field. It created the possibility of testing several sites at once. The results have already been exploited in a number of biotechnological applications. Modified enzymes are often present in powdered laundry detergents and cloned into microorganisms that industries use as factories to produce molecules that are difficult to synthesize.

The process still relies on (educated) trial and error, and we certainly do not have the same confidence in our redesign that, for example, a mechanical engineer does, but advances in chemistry, physics, and molecular and computational biology are converging to give us the ability to control, and predict, the functional roles of molecular sites. This development is having a strong impact on nanotechnology.

Another recently achieved result is the modification of proteins such that the cell that contains them "learns" how to incorporate unnatural amino acids into newly synthesized molecules, which creates the possibility of a much wider chemical repertoire.

In this rapidly evolving field, we cannot predict what will happen next and which computational tools will become necessary. However, we need methods to dissect the role of specific amino acids or regions within protein molecules, and we must also develop computational methods to optimize the design of combinatorial experiments, which are still rather limited.

Suggested Reading

Brannigan, J.A. and Wilkinson, A.J. Protein engineering 20 years on, *Natl. Rev. Mol. Cell. Biol.* 3, 964-970, 2002.

Yano, J.K. and Poulos, T.L. New understandings of thermostable and peizostable enzymes, *Curr. Opin. Biotechnol.* 14, 360–365, 2003.

Fernandez-Gacio, A., Uguen, M., and Fastrez J. Phage display as a tool for the directed evolution of enzymes, *Trends Biotechnol.* 21, 408–414, 2003.

Friboulet, A., Izadyar, L., Avalle, B., Roseto, A., and Thomas, D. Abzyme generation using an anti-idiotypic antibody as the "internal image" of an enzyme active site, *Appl. Biochem. Biotechnol.* 47, 229–237, 1994.

Lerner, R.A. and Janda, K.D. Catalytic antibodies: Evolution of protein function in real time, *EXS* 73, 121–138, 1995.

Failla, C., Pizzi, E., De Francesco, R., and Tramontano, A. Redesigning the substrate specificity of the hepatitis C virus NS3 protease, *Fold Des.* 1, 35–42, 1996.

Tramontano, A., Gololobov, G., and Paul, S. Proteolytic antibodies: Origins, selection and induction, *Chem. Immunol.* 77, 1–17, 2000.

Hedstrom, L. Serine protease mechanism and specificity, *Chem. Rev.* 102, 4501–4524, 2002.

Fersht, A.R., Matouschek, A., and Serrano L. The folding of an enzyme. I. Theory of protein engineering analysis of stability and pathway of protein folding, *J. Mol. Biol.* 224, 771–782, 1992.

Oliveberg, M. Characterisation of the transition states for protein folding: Towards a new level of mechanistic detail in protein engineering analysis, *Curr. Opin. Struct. Biol.* 11, 94–100, 2001.

Conclusions

The list of problems discussed in this book is only partial. Not even all the classical aspects of protein bioinformatics have been treated at a sufficient level of detail. Still, none of these problems is completely and satisfactorily solved. In some cases, the robustness of existing methods is challenged by the deluge of data. In other cases, the complexity of the task is increasing, rather than decreasing, with time. These challenges and complexities are providing an explosion of opportunities in bioinformatics.

The emergence of genomics and postgenomics, coupled with the rise in computing capabilities, has produced a shift of paradigm in biology from hypothesis-driven to data-driven approaches. In hypothesis-driven research, the goal is to understand the properties of a particular state of a particular system by successively decomposing it into subsystems until its behavior can be appropriately understood. Data-driven research relies on the study of a set, as complete as possible, of basic components of complex biological systems and integration of the basic components into a system-wide model.

The power of modern biology lies in the combination of these two approaches, but such a combination requires a high level of integration of the data and of sophistication of the tools. This requirement is the challenge that computational biologists and bioinformaticians must meet. They do not have an easy task, because they must blend computer science, software engineering, statistics, and biology. On the other hand, the strength of this discipline is exactly its ability to attract investigators from several fields and incorporate innovative ideas developed in different areas of science.

The purpose of this book is to provide newcomers to the field, as well as students, with a roadmap of the techniques that have been proposed to solve many problems, and the strengths and weaknesses of these techniques are highlighted.

Many more problems are not treated or are only hinted at in this book. They include, for example, the reconstruction of metabolic networks and of protein interaction maps, as well as the modeling of entire cells. Undoubtedly, more such problems will appear in the near future. However, I am confident that some of the aspects emphasized in this book will remain of paramount importance, such as the evaluation of the appropriateness of the approximations, the accurate testing of the reliability of methods, and the importance of being aware of the limitations of the data produced by experimental methods.

Index

A

Abstracts, use of in protein function prediction, 49–50

Active site detection, in functional site identification, 112–113

Affinity chromatography experiment, protein–protein interaction, 129

Antigen, antibody bound to, 124

Antiparallel helices, parallel helices, packing against each other, protein design, 153

Automatic methods in protein design, 155–160

B

Bidimensional gels, use in protein function prediction, 64

Binding sites in protein engineering, 166–167

Biological databases, searching of, 24–28

Biological function
defined, 45–46
dimensions of, 45

Boltzmann equation, structure prediction, energetic calculation, 72–76

C

Catalytic antibodies, in protein engineering, 167–170

Catalyzed biochemical reaction, in protein engineering, 170

Combinatorial design, in protein engineering, 170–172

Combining functions, in protein engineering, 164

Computational linguistics techniques, in protein function prediction, 50

Connecting neurons, in protein feature prediction from sequence, 35

Cysteine, hydrophobicity scale use, 97

D

DALI algorithm, 109–111

Deterministic patterns in protein feature prediction from sequence, 31–33

Diagnostic patterns in microarrays, detection of, 61–62

Dictionary extracted from databases, in protein function prediction, 49

Direct/inverse folding problem, protein design, 150

Distance geometry-based methods, for protein–small molecule interaction, 143–144

Disulfide bridge, in protein engineering, 166

Docking
application to Fourier transforms, 135
representation of protein structures for, 131–136
computational approaches, 131–133
conformational space search, 133–134
scoring docking solutions, 134–136

*For Product Safety Concerns and Information please contact
our EU representative GPSR@taylorandfrancis.com Taylor & Francis
Verlag GmbH, Kaufingerstraße 24, 80331 München, Germany*

T - #0028 - 160425 - C8 - 234/156/12 [14] - CB - 9781584884910 - Gloss Lamination